WHAT YOUR COLLEAGUES ARE SAYING . . .

Vanessa is a rock star—both literally and as a math educator—and her book will help you understand the origins of math trauma and its practical solutions. Now is the time and this is the book to help your students develop a new relationship with mathematics and create a new story about themselves as mathematicians.

Dan Myer
Director of Research, Amplify
Oakland, CA

In Math Therapy™, *Vakharia delivers a powerful message wrapped in warmth and practicality. As someone deeply entrenched in the world of mathematics, I appreciate how she seamlessly weaves personal anecdotes with actionable advice. Her call for educators to embrace the role of math therapists resonates deeply with the ICUCARE framework, making this book not just important but essential reading for anyone invested in equitable math education.*

Pamela Seda
Founder and CEO, Seda Educational Consulting, LLC
Coauthor, *Choosing to See: A Framework for Equity in the Math Classroom*
Atlanta, GA

In her rousing and radiantly witty style, Vanessa Vakharia offers extremely practical advice for schoolteachers willing to accept that a student's emotional relationship with math is vitally important.

Francis Su
Author, *Mathematics for Human Flourishing*
Pasadena, CA

This book is a gift to society. For too long, many people have believed that they are not "a math person" when the reality is that this belief says much more about how they learned math than about themself. Vanessa's book helps normalize these feelings, unpacks why the trauma has happened, presents a plan for healing the trauma, and paints a picture of what's possible. This is the perfect book for anyone who thinks that they're not good at math or don't belong in mathematics.

Robert Kaplinsky
President of Grassroots Workshops
Long Beach, CA

Vakharia has already been sounding the alarm about the very real trauma that so many experience in math class. Now she's put her therapeutic know-how into Math Therapy™, *a book to help cure math trauma—or better yet, not inflict it in the first place!*

Dan Finkel
Founder, Math for Love
Seattle, WA

I feel seen, as I know many will, in this book and WOW . . . I can read it. I'm tired of heavy academic language just to sound smart. Vanessa Vakharia is smart AND I can follow what's going on!! I love it.

Crystal M. Watson
Educator, Cincinnati Public Schools
Crystal M. Watson Consulting
Cincinnati, OH

Vakharia makes math an inclusive adventure. Drawing from her own experiences, she offers tangible, transformative steps to mend our uneasy relationship with math. Told in an unforgettable voice, this timely book is a beacon for educators committed to revolutionizing the math experience for all.

Anthony Bonato
Professor of Mathematics,
Toronto Metropolitan University
Toronto, Ontario, Canada

If you are interested in your students examining and rebuilding their existing relationships with mathematics, then this book is for you! Vanessa offers a clear path for educators to help motivate their students to makeover their connection to math and math class. Not only will you know how to better help your students build a positive math identity, but you'll also have the best time engaging with this joy-filled book!

Zak Champagne
Teacher, The Discovery School
Jacksonville, FL

Math Therapy™ *is the book many educators didn't know they needed. The words of inspiration and practical strategies between these pages will change lives. Whether you believe you are a "math person" or not, this book will be a game-changer to help you heal and help others around you heal. It will become a guide for rewriting our collective math narrative.*

Deborah Peart
CEO & Queen Mather, Mathematical Mind
Ocala, FL

As someone who hated math when I was a student and during my first few years as a teacher, I can attest to the power of addressing past traumas and developing a fresh and healthy relationship with mathematics. Vanessa's book is full of useful strategies and tips to help you shift your mindset about math and give you more confidence as a classroom teacher. It's a must-read for any teacher who has a less-than-favorable view of mathematics.

Mike Flynn
Author and Chief Executive Officer, Flynn Education, Inc.
Northampton, MA

Math class is so much more than teaching math. I have learned a great deal about my students by asking about their math stories, which for some include anxiety from past math experiences. In this book, Vanessa aligns our values to concrete strategies to help everyone see that they are a math person.

Howie Hua
Math Lecturer, Fresno State
Fresno, CA

She is not called "The Math Guru" for no reason! Vakharia does a fabulous job of providing specific examples of how to address math trauma, as well as activities and examples of what teachers can do each and every day to ensure this doesn't continue to happen in our classrooms. Her quick wit and strong voice make you feel that you are right there having a conversation with her, and this is a book you will not be able to put down!

Kristin Kanaskie Grotewold
K–5 Math Coach, Iowa Council of Teachers of Mathematics
Urbandale, IA

This book is a healing gift to math education. Vanessa walks you through how to help students (and yourself) heal their relationship with math. The stories alone are a masterclass in inviting students to (metaphorically) sit on her pink velvet couch and transform the way they see themselves in relation to math, and beyond.

Liesl McConchie
Math With the Brain in Mind
San Diego, CA

This book is a valuable and engaging resource for mathematics teachers to counter the pervasive mathematics trauma that exists within so many in our society. By healing, students can come to see that we are all mathematics people and that the ability to excel in this discipline lies inside them.

Lidia Gonzalez
Professor, Department of Mathematics and Computer Science,
York College, City University of New York (CUNY)
New York, NY

Math Therapy™ *provides teachers with the tools to humanize their classrooms and help all students rediscover their inherent connection to mathematics. Educators reading this book will discover tools and strategies to talk to young mathematicians in a way that empowers them to take back their mathematical identity. Ultimately, everyone will feel validated by following the strategies of* Math Therapy™.

Vanessa's book sheds light on the anxiety present in many math classrooms and provides concrete and easy-to-implement strategies for engaging students in mathematical reflections and building a positive math identity. Everyone who teaches math, at any grade, should read this book.

This book fights back against the notion that there's such a thing as a "math person" and challenges readers to eliminate any sort of limiting beliefs we have about ourselves. Vakharia breaks down everything wrong with the education system. Every educator needs to pick up this book to learn how we can repair this next generation of humans' relationship with math!

Math Therapy™ *is more than a guide to mathematical competence—it's a transformative journey toward emotional well-being and intellectual freedom. Vakharia demystifies math, turning it from a daunting subject into a path of discovery and growth. Essential for educators and anyone eager to embrace math's beauty without fear or frustration.*

MA±H THERAPY™

MA±H THERAPY™

5 STEPS TO HELP YOUR STUDENTS OVERCOME MATH TRAUMA AND BUILD A BETTER RELATIONSHIP WITH MATH

VANESSA "THE MATH GURU" VAKHARIA

FOREWORD BY JO BOALER

CORWIN Mathematics

FOR INFORMATION:

Corwin

A Sage Company

2455 Teller Road

Thousand Oaks, California 91320

(800) 233-9936

www.corwin.com

Sage Publications Ltd.

1 Oliver's Yard

55 City Road

London EC1Y 1SP

United Kingdom

Sage Publications India Pvt. Ltd.

Unit No 323-333, Third Floor, F-Block

International Trade Tower Nehru Place

New Delhi 110 019

India

Sage Publications Asia-Pacific Pte. Ltd.

18 Cross Street #10-10/11/12

China Square Central

Singapore 048423

Vice President and
    Editorial Director: Monica Eckman

Senior Acquisitions
    Editor, STEM: Debbie Hardin

Senior Editorial Assistant: Nyle De Leon

Production Editor: Tori Mirsadjadi

Copy Editor: Amy Hanquist Harris

Typesetter: C&M Digitals (P) Ltd.

Proofreader: Heather Kerrigan

Indexer: Integra

Designer: Scott Van Atta

Marketing Manager: Margaret O'Connor

Printed in the United States of America

Library of Congress Control Number: 2024019735

This book is printed on acid-free paper.

24 25 26 27 28 10 9 8 7 6 5 4 3 2 1

CONTENTS

Visit the author's website at
maththerapy.com
for downloadable resources.

FOREWORD

by Jo Boaler

Nomellini and Olivier Professor
in the Graduate School of Education
(Mathematics), Stanford University

When Vanessa asked me to write the foreword for this book, it only took me a nanosecond to say yes. Why? Because Vanessa is my good friend and because she is helping all of you reading this book to conquer math trauma and anxiety—your own and your students', a mission that she and I are passionate about. Math anxiety is a real condition that robs people of pleasure and limits their life chances, and it is staggeringly widespread, as Vanessa describes in the book.

Neuroscientists have shown that, when people with math anxiety see numbers, a fear center lights up in their brain. That is the same fear center that lights up when we see snakes and spiders.[1] How great is it, then, that we now have a math therapist, also known as the Lady Gaga of math education, giving us practical strategies to rid us of anxiety and improve our own and our students' mathematical relationships?

Every year, I teach undergraduate students at Stanford, and even among this high-achieving group of young people, I meet many who have significant mathematics damage, even trauma. Some students have told me they are so afraid of numbers they won't use credit cards. This is because they have come from a school system that promotes the idea that you need a special "gift" to be good at math and that the gift is usually reserved for certain people. When these faulty ideas are combined with a system of teaching mathematics as a set of procedures that you must memorize and reproduce at speed, you have the perfect conditions to create math anxiety. I have been working for years to combat this math anxiety myself. Over a 10-week course I teach, the students change—they start to see mathematics

---

[1]Adelson, R. (2014). Nervous about numbers. *APS Observer*, 27; Young, C. B., Wu, S. S., & Menon, V. (2012). The neurodevelopmental basis of math anxiety. *Psychological Science*, 23(5), 492–501.

differently, and they start to see themselves differently.[2] Vanessa and I know this change is possible, and we have both seen that it can create different lives for people. It is no wonder we are both so passionate about sharing the ideas in this book.

Vanessa and I agree that math fear also comes, in part, from the myth that being good at math means you are "smart." Personally, I think that being a stand-up comic or a poet is much harder than factoring a quadratic, but the idea contributes to the pressure students feel. I also believe that people only think math is so difficult and inaccessible because they have experienced bad teaching. I love the story Vanessa tells in this book of being set free by the teacher she had in 12th grade. All of us as educators have the power to make that change for students.

As Vanessa points out in this book, we need to give people the opportunity to develop different mindsets, and that means changes all through the school system, from teaching ideas to testing and grouping. Many educators feel helpless in changing students' relationships inside a system that gives students the wrong messages at every turn, but Vanessa highlights the changes that can and should be made by classroom teachers—and they are all about changing the culture of classrooms. This is backed up by research. The latest research on ways to encourage a growth mindset shows that mindsets do not change when we just share words with students; they change when we create mindset cultures.[3] And the secrets of these mindset cultures are set out in this book.

Mindset is so important, as it is all about our ability to change. Research studies have found that when people learn about mindset they realize they are able to change in other ways, and they learn that all people are able to change—not only in terms of learning, but in their behavior and attitudes. This is why mindset interventions have helped students dial back on aggression and racism and improve their health, among other changes, as well as experience transformative changes in learning.[4] But as with almost anything these days, this idea is not without controversy. There are people pushing back on the importance of mindset—and some are belligerent. I really think that it is impossible to bring forward anything positive and impactful without having a vocal minority that disagree, often aggressively. Often, such

---

[2]Boaler, J. (2024, January). *"If only I had known": Messages for students who would like to learn mathematics to high levels.* youcubed. https://www.youcubed .org/wp-content/uploads/2024/01/Student-Letter-2024.pdf

[3]youcubed. (n.d.). *Mindset evidence: What is all the controversy about mindset?* https://www.youcubed.org/resource/mindset-evidence/

[4]Yeager, D. S., Trzesniewski, K. H., & Dweck, C. S. (2013). An implicit theories of personality intervention reduces adolescent aggression in response to victimization and exclusion. *Child Development, 84*(3), 970–988; Zahrt, O. H., & Crum, A. J. (2017). Perceived physical activity and mortality: Evidence from three nationally representative U.S. samples. *Health Psychology, 36*(11), 1017–1025.

detractors reject the idea that mindset is effective because they base their ideas on cases when people have only shared mindset ideas without changing the culture. Now we have new research that shows that mindset is extremely important, for learning and for life, when we do more than change words.[5] Vanessa's book shows educators how to make these changes in culture.

I love all the chapters in Vanessa's book, but a particular favorite was the chapter on student motivation. As someone who has trained teachers for decades (in Stanford's teacher education program), I cannot tell you how many times I have been asked by new teachers what they should do with "unmotivated students." When I am asked that question, I always share something I heard a great teacher say. The teacher was Carlos Cabana, and one day, when I was visiting his classroom, I overheard his conversation with a student teacher. The student teacher asked what he should do with the "unmotivated students in the class" who were misbehaving. I may not be remembering the exact words Carlos used, but this is my memory of what he said:

> I have never met a student who is not motivated to learn. I have
> met students who have constructed layers of ice around themselves
> that are so thick and so strong, it takes a long time to break
> through them. But that is my job.

There is so much in these statements, including Carlos's unwavering belief in the potential of all students. Students construct these layers of ice when they get the idea that they cannot be successful, and that is what we can change for students. To do that, we cannot be put off by the appearance of not caring or by their behaving badly. Those to me are signs that someone, somewhere in their lives has given the student the idea they can never be successful. And they have internalized the idea that it is better to act up than it is to fail.

We spend so much time in education worrying about knowledge and the way it is delivered to students—when students' beliefs and ideas about mathematics and themselves have so much more impact on their learning. Historically, we have barely given students' ideas and emotions any consideration in mathematics education. Vanessa's work is the antidote. *Math Therapy*™ is a handbook of ideas and strategies for giving your students mathematical confidence and the appreciation and self-love they need.

I invite you to learn from Vanessa's ideas and enjoy them. I smiled and laughed my way through the book, and I was even a little sad to come to the end. I particularly appreciated the many analogies. My personal favorite is

---

[5] youcubed. (n.d.). *Mindset evidence: What is all the controversy about mindset?* https://www.youcubed.org/resource/mindset-evidence/

the description of brain pathways being laid down as an experience of walking through deep, freshly fallen snow. Reading this book feels like having a conversation with a wise and caring friend, the one and only Lady Gaga of math education. Enjoy it, reflect on the words, and let Vanessa help you make real, actionable changes that can improve the lives of those you work with—and maybe even your own.

PREFACE

Anything Is Possible (Yes, Even Math!)

DEAR MATH-THERAPIST-IN-THE-MAKING

There is one job in this world that I truly believe has more power than any other to change the course of our future as a species, both on an individual and global level.

Who IS this powerful person I'm getting all fired up about? YOU! The most powerful person in the world is the math teacher. Yes, I may be biased, but hear me out!

I believe that every single math teacher has an opportunity that *no other teacher has*. I'm not hating on the history and geography teachers—I love you guys! But here's the thing: By early to mid-elementary school, many of our students have gotten the message that there is this THING on the planet called math that certain people can—and can't—do (Hachey, 2021; Levine & Pantoja, 2021), and that belief takes root inside of them, more often than not, laying the foundation for a lifetime of limiting beliefs that start stacking up. The next time they face an obstacle, they might say to themselves, "Is this just another thing I was born unable to do?" Not making the soccer team turns into, "I guess I'm just not athletic." Getting passed over for a job turns into, "I'm unemployable—why bother trying?" Getting ghosted after a first date turns into, "I'm unlovable and destined to be single forever."

And on it goes. Whether or not a child is given permission to dream big or small often starts and ends with whether or not they are given the message that they are a "math person" . . . or not. That's why it is YOU who has the incredible power and privilege of changing their lives far beyond your class-room walls. It is YOU who holds the keys to unlocking their mathematical mindsets, releasing the dreams that lie within. And that's why I wrote this book. Math Therapy isn't just about helping students kick butt in math class; what it's really about—and what I believe *education* is really about—is empowering students to develop the skills they need to live a life in which they truly believe that *anything is possible*, even a better relationship with math! And I want to help you *help your students* do just that.

I am not asking you to believe that every single child is capable of achieving an A+ in your class, of being a rocket scientist, or even of performing mental calculations with total accuracy. I am asking you to believe—really believe—that every child is capable of *building a better relationship with math than they have right now*. Our students *know* whether we believe in them or not, and that in turn shapes whether or not they ultimately believe *in themselves*. I want you to imagine that you are a parent watching a child (caveat: *without* a physical disability) learn how to walk. No matter how long it takes for them to take their first steps, no matter how many times they fall in the process, and no matter how much they insist that walking is just too hard, I'm going to guess that you would never once pause to think to yourself, Hmmm . . . maybe little Johnny just isn't a "walking person"; maybe walking just isn't his thing—that's cool, he has other strengths. Am I right? You would coax and coach and cheer on little Johnny with ALL of the belief in your heart that, with patience and perseverance, little Johnny was going to be an amazing walker! It is THAT kind of belief I want you

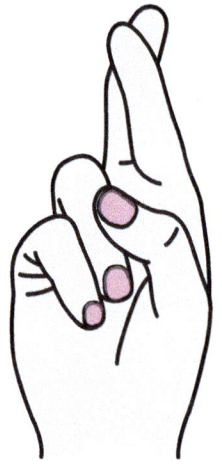

to bring to the table when it comes to your students' relationships with math. As hard as this is to hear, in many cases you might be the only adult in some of your students' lives who actually believes in their capacity to grow and change, both inside, and sometimes even outside, of math class. If THEY can sense that you believe they can make progress, they, too, will begin to believe it as well. Channel your inner Ted Lasso and B-E-L-I-E-V-E!

THE "OFFICIAL" PREFACE

Hello. My name is Vanessa Vakharia, and I failed Grade 11 math twice.

(Actually, almost three times—but who's counting?!)

Let me back up. I went to a fairly academic school in Toronto, which shall remain unnamed. Let's just say that their mantra was "We're not a high school. We're a *collegiate institute*," which basically translated to "If you're not getting an A, we really don't want you around to give us a bad rep." Well, I was very busy doing the latter by insisting that my only interests included (a) becoming a rock star and (b) marrying Keanu Reeves, and I certainly wasn't getting an A in math class (or any class for that matter, except for art). I was the "weird" kid who wore blue lipstick, smoked cigarettes, and whose report comments seemed to always contain some variation of "Vanessa is not living up to her full potential." While all of this was going down, it seemed like almost everyone in my life was reaffirming my life choices by letting me know, in one way or another, that failing math was my destiny—girls like me weren't made for math: *I just wasn't a math person.*

Here's the thing: I totally agreed. It was the 1990s. Math was for losers, and I was meant for the Hollywood Hills. That was a no-brainer. But you know who (thankfully) didn't agree? My parents.

After I *finally* passed math in summer school, my parents decided it was time for a change. So off I went to the private "alternative" school down the street, which was low-key known as the "school for misfits." It. Was. Heaven.

It was NOTHING like my old school. First of all, it was in an office building. We had to take ELEVATORS to get to our classrooms. Second, we called teachers by their FIRST NAMES, which, for a teenager, is just about the coolest thing you can do. Third, and most important, there were only about 100 kids in the entire school, which meant that there were no cliques. There were no "jocks" or "nerds" or "math people" or "non-math people"— we were all just "misfits" who, for one reason or another, found ourselves in this magical space. The thing is, none of our teachers thought we were there because we couldn't hack regular school or because we were stupid—it was the opposite. They all believed that we were there because we were *smart* and that our inner light simply hadn't been given permission to shine through in a traditional educational setting. When I walked into my brand-new Grade 12 math class, I felt something I had never truly felt before: My teacher seemed to believe . . . *in me.*

I don't remember the exact details, but my first encounter with Ewa, my math teacher (and the teacher who would literally change the trajectory of my entire *life*), went a little something like this:

Me:　Hey, hi, just wanted to let you know that I'm not so sure how this is going to go because I'm not a math person.

Ewa:　Pardon me?

Me:　I said, I'M NOT A MATH PERSON.

Ewa:　(long, dramatic pause) That's. Not. A. Thing.

Everything changed after that. It's not like—POOF—I suddenly became a math genius or something, but my *relationship* with math changed. Like, TOTALLY changed. I started doing things I had NEVER done before. For example, when I didn't understand something, I raised my hand and asked for clarification. If something didn't make sense, I asked Ewa to explain it to me a different way. I took notes and did my homework and checked my answers at the back of the book, and when I didn't get something right, I sought out help because I believed that I had the *capacity* to understand it. I discovered that I loved the methodical nature of whittling complex algebraic equations down to a simplified expression; I reveled in the art of mathematical proofs, in the neatness of setting up and then solving a problem; and most of all, I found the unparalleled gratification of helping others discover their own ability to do something that they had been convinced they would never, ever be capable of doing. Check out Figures P.1 and P.2 for photos of Ewa and me back in the day and Figure P.3 for a more recent shot!

FIGURE P.1 VINTAGE HIGH SCHOOL YEARBOOK PHOTO OF EWA KASINSKA

FIGURE P.2 VINTAGE HIGH SCHOOL YEARBOOK PHOTO OF ME

Ewa Kasinska

SOURCE: Hudson Taylor

SOURCE: Hudson Taylor

● I went on to be Ewa's student teacher in teacher's college, she edited my first math book series for kids, and we are still bffs to this day!

SOURCE: Ewa Kasinska

Year ____-__	Semester: 2		
Course: **OAC Finite Mathematics**	Code: **MFNOA**	Teacher: **Ewa M. Kasinska**	
Applications of matrix algebra to solving systems of equations and inequalities, combinatorics, probability, and applications.			

A truly breathtaking performance!

Absents: **2 / 92**	Lates: **2 / 92**	Grade: **96**	Class Avg: **59.1**	____: **0.00**

I ended up with a 96% in Ewa's math class. I know, I know—it's not about the mark! But, guys . . . come on. It was a BIG deal. But what was a way BIGGER deal was that I had gone from "There are some things on this planet that I just can't do" straight to "If I can get a 96% in math class after being told for most of my life that *I literally can't*, WHAT ELSE CAN I DO IN THIS WORLD?!"

My entire life changed.

I hadn't realized it, but the belief that math was this THING that I just couldn't do had been holding me back in SO many ways. Years later, I discovered that this is known as a *limiting belief* (Heston, 2015). Limiting beliefs are essentially a type of negative thought that holds people back by creating impenetrable mental barriers that prevent them from moving forward with confidence. That was exACTLY what had happened to me until I met Ewa. The limiting belief that there was just something that I couldn't do (math!) had subconsciously been in the back of my head any time I considered trying something new or dreaming my biggest dreams. It was like I was always waiting for the other shoe to drop, always waiting to discover what ELSE I was just innately incapable of doing. But after Ewa, everything was different. I wanted to try EVERYTHING. I wanted to show EVERY SINGLE PERSON on the planet that they had been sold a big fat lie, that there was no such thing as a "math person," that there was no such thing as any KIND of person. I wanted to go after all of my crazy wild rock star dreams. I wanted to . . . start a band.

(Okay so, *spoiler alert*: I DID end up starting a band, and if you're asking yourself, What's the connection between healing your math trauma and being on stage in front of over 10,000 people opening for Bon Jovi, you're in luck—because you're eventually going to find out!)

What started as a somewhat personal vendetta against a system that had tried to shut me down turned into a very personal mission to show literally EVERYONE that they, too, could build a better relationship with math than the one they'd been sentenced to and that they, too, could rediscover what it was like to live a life of wondrous, boundless *hope*.

Now, okay, I know it might sound a little out there that I'm suggesting that building a better relationship with math might be basically like the gateway drug to building a life permeated by hope, but hear me out:

- *Most* kids are told, in one way or another, that they have a deficit when it comes to math and have carried that belief into adulthood (Sun, 2014).

- *Most* kids are told, early on in their lives, that *they are simply not that good at math* and should focus on their actual "strengths" will carry that belief with them for the rest of their lives.

- *Most* kids get the message that they just don't have what it takes, period.

Think about that for a second.

By the age of 6, 50% of our students are already exhibiting signs of math anxiety (Sokolowski & Ansari, 2017), and many have gotten the message that math is something on this planet that they are simply born *incapable* of doing. This belief that they innately can't do math is the scaffolding upon which many of them build walls constructed and obstructed by limiting beliefs. Maybe some of you even got this same message and built up those same walls—you already know that I did! Being told that math "isn't for us" or that we're "not math people" or that we should "forget about math in favor of our other strengths" are some of the first experiences many of us have that suggest that there are *skills on this planet that are unavailable to us*.

Because math ability is so often culturally associated with intelligence, being told that we can't do math has the painful effect of making us feel less-than, affecting our self-worth in all areas of our lives, both personal and professional. The culmination of experiences that lead us to the conclusion that we're innately incapable of math can be traumatizing, and most of us never fully move on. But it doesn't have to be that way.

ABOUT THIS BOOK

I wrote *Math Therapy* as a message of hope. It is the culmination of more than 20 years as a math educator, of hearing hundreds of math stories on my podcast and in interviews, and of helping thousands of students build a better relationship with math, and as a result, with themselves.

I was blessed with an incredible number of privileges that helped me build a better relationship with math: parents who cared, a private alternative school education, and a skilled teacher who believed in me. But your students, too, are blessed—because they have YOU.

I'm going to reiterate what I said at the start of this preface because it is THAT important: Math teachers have an opportunity to change the entire trajectory of a student's life by helping them heal their relationship with math.

Now, even if you're not an official teacher of math, we are all teachers of *math attitudes*. Regardless of what grade you teach, what role you play in education, or whether you consider yourself an educator at all, *your* relationship with math affects those around you. The way you do math, talk about doing math, or talk about *who* can do math has more of an impact than you might think. It teaches those around us how to feel about math, either directly or indirectly. That's why it's important for ALL of us to build a better relationship with math, and that's why no matter who you are, the fact that YOU are about to learn all about how to heal from math trauma is, in itself, a gift that you will inadvertently regift to everyone you know.

ABOUT LANGUAGE

Let's talk language. I have done my best to write in a way that feels the most accessible—and most authentic—to me (and hopefully you!), all the while being conscious that language is ever-evolving. In my work, when I want to write, teach, and speak in a way that is relatable to educators, students, and anyone out there who wants to build a better relationship with math, I don't think that I can do it by using the pedantic language required by most academic institutions. I have often been underestimated, overlooked, and not taken seriously because of the way I speak. I have been told that I should say "like" less often, fix my "Valley girl" accent (I'm from Toronto, btw, so like . . . very far from any peak or valley), and "be more serious." Sorry not sorry—that's just not who I am. I wrote this book from my heart to yours, and I hope that you hear my voice in every word that I have written.

This is why I use ALL CAPS. A. Lot. And why I use *italics* and exclamation points liberally!

Language is constructed and in making language choices. I recognize that we are often complicit in its construction, which is why I want to acknowledge certain choices that I made. When I am writing about racialized people, I use the current most recognized term for the group—for example, *Black*, *Latine*, and *Indigenous*. I made the choice not to capitalize "white" when I am referring to race because that can be offensive to some. I use the phrase *people of color* to describe all those who are non-white. I use the phrase *underrepresented groups in mathematics* in keeping with the definition of the National Science Foundation to describe those whose representation in mathematics degree programs and math-related careers is lower than their representation in the U.S. population (National Science Foundation [NSF], n.d.). When I speak of "marginalized" students, I do so to describe those who additionally experience systemic disadvantages such as those related to race, ethnicity, socioeconomic status, gender, sexual orientation, disability, or other characteristics. When using terms such as *men*, *women*, *girls*, and *boys*, I refer to all who identify with these terms. I know that some do not identify with binary genders, and I have avoided using *he* or *she* when a person's preferred designation is not known, choosing instead to use non-gendered terms such as *students*, *individuals*, and *people* or the pronoun *they*. I admit that I use "you guys" at times. I don't literally mean *guys*! But "y'all" just doesn't roll off my tongue! All this is to say that I know that my language choices aren't perfect, but I hope that this at least begins to clarify why I made the choices that I did. I also want to add that when I'm telling stories throughout these pages (which I do, a LOT!) that, in some cases, names of people have been changed, and in other cases (aka where permission has been granted), they have not.

Now, let's talk about the f*#king elephant in the room: swearing. If you're familiar with my podcast or if you know me personally, you know that I swear—a lot. I know that salty language isn't everyone's cup of tea, but I also wanted to write this book in my authentic voice, so as a compromise, I have bleeped out a few letters to keep things PG!

Finally, I am Canadian. Before I started writing this book, I don't think I ever noticed how many language differences there are between us and America. I mean, I feel like we're the only ones who even call it "America"?!

We say "100%"; they say "4.0." We say "mark"; they say "grade." We say "colOUr"; they say "color." And on it goes. I have tried to make my language as universal as possible in this book, but if I went all Canadian on you somewhere in these pages, just blame the maple syrup!

SO WHY NOW? AND WHY YOU?

As Hillel the Elder is attributed with saying, "If not you, then who? If not now, when?" Those words come to mind now, as we are at a moment in time in which we have never needed math more, yet we have never been more divided when it comes to who can actually access that math. Headlines continuously warn us both that the majority of jobs in the next decade will require math and yet that the majority of students around the world are failing to meet math standards (Mervosh & Wu, 2022). As educators, we understand that behind those clickbait numbers are actual, real kids who are suffering; these are young people who don't *coincidentally* happen to all be bad at math or to all be simply unmotivated or lazy or incapable of learning. But they do have something in common. These numbers represent children who, in one way or another, have likely experienced math trauma—a concept that has been mentioned in passing by many but that has not been properly defined, explored, unpacked, and addressed.

As I will discuss in Chapter 1, *most* of us have experienced math trauma *at least* once, but those from communities that are traditionally (and currently) marginalized in math education are even more likely to be exposed (Duncombe, 2021; Gonzalez, 2023; Louie, 2017) to scholastic practices that might lead to math trauma, making it even harder for them to feel a sense of belonging in the world of math. When we neglect to heal students' relationships with math and we neglect to provide equitable access to education, we elect to maintain a status quo in which we include some—but ultimately exclude most.

I know that some of you are likely wondering how you can carry out this work when the mere mention of emotions in math class can be the cause for alarm at the school, district, or even state level. Or when you have so much content to get through that it seems like math trauma can't possibly make it onto your to-do list. I totally get it. Here's what brain-based educator Liesl McConchie says: "A student's emotional relationship with math is foundational to their cognitive relationship with math." This is what I tell anyone who's skeptical that helping a student build a better relationship with math will more likely than not lead to improved proficiency in actual math skills. Seriously. You can have the best content and pedagogy in the world, but if a student has shut down and built walls, it's never going to reach them. THAT is why Math Therapy is a crucial first step in ANY sort of initiative directed at improving

math performance and proficiency, and hopefully this is helpful for you if you feel like you have to justify your commitment to this practice.

If you picked up this book, you are likely as worried about our collective relationship with math as I am. Maybe you're worried that your students won't be able to get into college or into a career trajectory that will enable them to support themselves; perhaps you're worried that they won't have the financial literacy needed to make wise decisions and live within their means; maybe you're nervous that they won't have the foundational math skills needed to decipher the proliferation of numbers and stats that permeate our screens, and as a result, they'll get sucked into some whack conspiracy theory and, like, join a cult; or maybe you're stressed out about the idea that a large percentage of today's students will be working in jobs that don't even exist today and that they simply won't be prepared with the skills needed to access those jobs. All of those concerns are totally valid and totally warranted. But you guys know where I stand: The one thing that worries me most, above all else, is that our students might grow up believing that they are undeserving of going after their wildest dreams. That, to me, is unacceptable. And that, to me, starts with math.

Every single child should have the privilege of believing that *anything is possible*. And in fact, *we* as a society, NEED them to believe that for our OWN benefit AS A SPECIES. I know it sounds extreme, but, guys, we NEED kids to grow up believing that there are solutions to the unsolved mysteries of our time! We NEED them to believe that climate change can be mitigated, that racism can be extinguished, that cancer can be cured, that peace and love can one day reign supreme! **If not them, then who? If not now, when?!**

> *Every single child should have the privilege of believing that anything is possible.*

In this book, you will discover the most important and impactful steps you can take to help your students heal their math trauma and build a better relationship with math. You will learn science-backed strategies that *really work* and unpack toolkits full of Actions, Activities, and quick adjustments to your classroom practice that you can implement effectively *right away, right now.*

I wrote *Math Therapy* as a message of hope, and it is *my* hope that you will bring your own magic to its pages. It takes a village to change the culture of mathematics, and it takes a leap of faith to believe that we can. I am so, so happy you're here.

yours in peace, love + pi
XOXO Vanessa

ACKNOWLEDGMENTS

I have so many people to thank that I am officially too overwhelmed by the prospect of leaving one of them out to even begin to list their names. So instead, I will say this: If you have ever believed in me, stood up for me, listened while I've ranted, laughed at my jokes, put in a good word for me, had my back, given me a pep talk, given me a reality check, helped me take care of myself, trusted me with a secret, given me a reason to trust you with one of mine, taught me literally anything, allowed me to read your birth chart, invited me to be a part of your journey on this planet, and/or—most importantly—given me the freedom to be myself, thank you. This book would not exist without you. ♡

To my Gemini twin, publishing soulmate, and dream editor, Debbie Hardin: there is no one else in the world I could have created this book with. Thank you for honoring my ALL-CAPS voice and for encouraging me never ever be too scared to use it. ILY.

PUBLISHER'S ACKNOWLEDGMENTS

Corwin gratefully acknowledges the contributions of the following reviewers:

Anthony Bonato
Professor, Toronto Metropolitan University
Toronto, Ontario, Canada

Karina Cousins
5th Grade Teacher
Cofounder, Learning Through Math, Palm Beach School District
Lake Worth, FL

Pooja Desai
Middle School Math Teacher, Trinity School
Brooklyn, NY

Ana Maria Estela
K–5 Math Coordinator, Trinity School
New York, NY

Lidia Gonzalez
Professor, Department of Mathematics and Computer Science, York College,
City University of New York
New York, NY

Kristin Kanaskie Grotewold
K–5 Math Coach, Iowa Council of Teachers of Mathematics
Urbandale, IA

Robert Kaplinsky
President, Grassroots Workshops
Long Beach, CA

Liesl McConchie
Owner, Math With the Brain in Mind
San Diego, CA

Katie McCormick
Math Teacher, Gahanna Middle School East
Centerburg, OH

Jamie Mitchell
Secondary Math Teacher, Halton District School Board
Brantford, Ontario, Canada

Leslie A. Mohlman
Secondary Math Teacher, Alpine School District
Lehi, UT

Deborah Peart
CEO & Queen Mather, My Mathematical Mind
Ocala, FL

Laura Vizdos Tomas
Math Coach, School District of Palm Beach County
Cofounder, LearningThroughMath.com
West Palm Beach, FL

Crystal M. Watson
Educator, Cincinnati Public Schools
Consultant, Crystal M. Watson Consulting
Cincinnati, OH

ABOUT THE AUTHOR (AND ILLUSTRATOR!)

Known as the Lady Gaga of math education, Vanessa Vakharia is the founder and director of The Math Guru, a super-cool boutique math and science tutoring studio in Toronto that's changing stereotypes about what math education looks like. She is also the host of the *Math Therapy* podcast, author of the *Math Hacks* Scholastic book series, math therapist, and lead singer/keytarist for the rock band Goodnight Sunrise. She has her bachelor of commerce degree, teaching degree, and masters of math education. She appears regularly on national television and news outlets as an expert in math education and math positivity, and her #goals are to be Oprah-level famous and to totally change math culture so that STEM is finally as cool as every single Taylor Swift song ever written. Vanessa failed Grade 11 math twice, which was the best thing that ever happened to her.

CREATE YOUR OWN MATHOGRAPHY

MATHOGRAPHY

noun

ma·thog·ra·phy

A mathography is like a math autobiography. Instead of focusing on your whole life, you talk about your journey with learning math, doing math, and literally anything math related. It's a way to reflect on your relationship with math and what math means to you!

The exercise: Tell me your math story!

That's it.

There are no rules, and I want you to just go for it, but I will say this: You may not realize it yet, but we each have a story that we tell ourselves every day. We have stories about who we are and how we got here, and we subconsciously live our lives by those stories. We might even hear those stories come out when we meet someone new and say something like this:

"Hi, my name is _____, and I love Keanu Reeves! I don't have any plants because I literally can't take care of living things, and I'm not great at cooking, so I eat out a lot. I like to do yoga, but I'm kind of unmotivated, so I end up bingeing reality TV most of the time. I love juicy convos, hanging out with friends, and also need my alone time because I'm a low-key introvert."

Notice how many things those few sentences tell us not just about this *mystery* person, but about how the person feels about THEMSELVES. Well, you can probably guess what I'm going to tell you next: We also all have stories about ourselves as math learners!

Now look, this isn't a huge exercise—it doesn't need to take hours, and I want you to have the freedom to do it in whichever modality feels authentic to you. If you want to take pen to paper, go for it. If you want to type it out, great. If you want to record a voice memo, amazing. If video confessionals are your thing, GET IT! If it makes you feel better, we're going to revisit this exercise later in the book (see Chapter 6), so you will have the chance to rethink and revise. Consider this a first draft—a low-pressure, low-key way for you to get a "before" snapshot of where you were when you started Math Therapy!

Oh—and you totally don't HAVE to do this—but I'm REALLY into journaling, and I think that having a dedicated space to write is SO important—for you AND your students! We're going to be talking about how to help your students

create their very own Math Therapy journal in Chapter 3, but now is a good time to pull out your math journal if you already have one, to buy a cute notebook to turn INTO your Math Therapy journal, or to download the journal pages I have created just for you (and your students) from the website I have created just for you at maththerapy.com!

Don't overthink it—just grab a cup of tea or whatever your drink of choice is, maybe light a candle, get in the zone, and just GO FOR IT. Remember, there are NO RULES. If you want to, send me your math story once you're done recording it, and either way, save it somewhere special so you can find it when it's time to revisit it. Ready . . . set . . . TELL ME YOUR MATH STORY!

CHAPTER 1

WHAT IS MATH TRAUMA ANYWAY?

In this chapter, we get to:

1. Figure out why so many of us are SO scared of math
2. Discuss the differences among stress, anxiety, and trauma
3. Dig into some of the main causes of math trauma
4. Explore what math trauma might look like in our classroom

Hello, math-therapists-in-the-making! I know we're all VERY excited to get right into it, and I promise that more than 200 pages of tools, strategies, templates, and resources await to help us on our way. Throughout these pages, you will be learning exactly how to turn your classroom into a Math Therapy classroom by weaving the principles of Math Therapy into both your physical space as well as every single part of your teaching practice so that you cultivate a culture in which students are consistently building a better relationship with math—and with themselves. But first, we need to understand why so many of our students (and teachers . . . cough, cough) have developed math trauma over the course of our lifetimes. And to do that, we need to understand what math trauma is in the first place. And to do THAT . . . we need to talk about anxiety, which is more often than not the way that math trauma manifests in our classrooms and irl.

WHY ARE WE SO SCARED OF MATH?

I'm going to go rogue and start this chapter with a question: Why is "math anxiety" a thing we hear about all the time, but we almost never hear the term *history anxiety* or *geography anxiety* or even *science anxiety*?! Seriously,

type "math anxiety" into the search engine of your choice, and your web browser will serve you countless articles about the topic, but type "history anxiety" or "science anxiety" into that same search bar, and you'll get articles titled "The History *of* Anxiety" or "The Science *of* Anxiety." Anxiety around another discipline is just not a thing in the same way it is with math. Math anxiety is so widespread that studies show that as many as 93% of Americans identify as having some form of math anxiety (West, 2022)! That number includes children, adults, students, *and* educators! Sometimes it almost feels like we're not even surprised that math anxiety is so commonplace; in fact, it's almost MORE surprising

when someone doesn't claim to be anxious about math. But have you ever stopped to wonder what it is about math that makes it so conducive to being anxiety-inducing for so many people?

Well, I have some theories.

Math Makes You Smart

As many of you know (because I won't shut up about it), I'm in a rock band (see Figure 1.1). We

FIGURE 1.1   ME ON STAGE AT LONDON MUSIC HALL

SOURCE: Dan Boshart

have played over 400 shows; we've honed our craft for over 10 years, working with vocal coaches, practicing our instruments, writing songs that get played on the radio—we're like, actual, legit musicians. Now, let me describe a scenario that happens over and over again at our gigs:

Picture me on stage (looking glam, obviously). I'm running around—JUMPING around—doing stage dives, singing at the top of my lungs on key while shredding sick solos on my keytar (which is basically a keyboard that you wear like a guitar with all sorts of effects on it that make it sound cool). I do this for 45 to 60 minutes straight during any given show, and at the end of our set, I usually get off stage and chat with the audience. Now, almost every time, someone will come up to me and be like, "Wow, you guys were so great, blah blah blah," and I'll be like, "Cool, thank you so much," and they'll say, "I know it's so tough being an artist; do you do this full time, or do you have another job?" and I'll say, "We definitely all have other jobs—I'm actually a math teacher," and then THEY will step back, shocked, and say, "OMG, WOW, YOU MUST BE SO SMART."

Let me repeat that: "You. Must. Be. So. SMART."

Here I am, shredding complicated keytar solos *while* singing IN TUNE *while* doing coordinated acrobatics, but none of THAT is enough to qualify me as "smart." The fact that I know math, however, is apparently the factor that seals my fate as being *intelligent*. What's that about?

Let's face it: We consider those who are "good at math" to be smarter than everyone else. We can likely all agree that Picasso and Mozart were incredibly talented individuals, but would we ever call them "smart?" That label tends to be reserved for the mathematically capable, cementing the idea that if you aren't good at math, then, well, you must be *un*intelligent, or worse—stupid. How. Stressful. Is. That?!

It's this classification of mathematics as *the* thing that makes someone smart—or not—that makes it so conducive to generating mass amounts of anxiety among most people. I mean, how stressful to think that if you don't understand a math concept right away then—BAM—you're not smart? Imagine the pressure! It is that pressure that ultimately generates a sense of anxiety around math. *Students aren't necessarily scared of math, of numbers, or of shapes; they're scared of being labeled as stupid.*

Math Makes You Money

There is nothing scarier to a parent than reading a headline like "STUDENTS ARE NOT MEETING OUR MATH STANDARDS" followed by a headline like "ALL FUTURE JOBS WILL REQUIRE MATHEMATICS SO WATCH OUT IF YOU DON'T WANT TO BE BROKE FOREVER." Fine, I'm

exaggerating—but only slightly. Seriously. Mass media are great at scaring the crap out of us.

For the past 10 years, I have been on the media list for math education in Canada. What that means is that any time there's a "math crisis" (e.g., standardized test scores drop by a single percentage point or more students are struggling in math class than they were last year or whatever the hot new math crisis might be), I'm the one that various media outlets will call for input. Every time this happens, I try to explain that the narrative that students need to get As in math in order to get anywhere in life is way more harmful than any of these manufactured crises, and furthermore, that narrative is possibly even *responsible* for the collapse of students' waning math abilities in the first place! I'm not going to pretend that grades don't matter—because they do—but I think we have to ask ourselves: At what cost?

HOT TIP

When your students start to fixate on getting the highest grade possible, help them pause and reflect on the "why." Do they need a specific grade for a certain reason? Are they just getting swept up on the hamster wheel of getting the highest mark possible? Are they comparing themselves to others? Focusing on the *why* doesn't eliminate the desire to achieve, but it does help students put things in perspective!

The idea that math ability correlates with your capacity to feed and shelter yourself is like, the most stressful thing I've literally ever heard, and I'm not even remotely surprised that, as a result, students and parents find math so stressful. I can't even tell you how many calls I get a week from parents looking for a math tutor for their kids because they're terrified that they will have NO future without an A in math class. *Students aren't scared of math; they're scared of having no hope for the future!*

Math Makes You Genetically Superior

Back in 2019, I was asked to be on a radio talk show to discuss a study that had proven that brain scans had shown there were no gender differences in math ability among children (Hunt, 2019). Cool. First of all, I would argue that there was literally no need for that study IN THE FIRST PLACE, as we already KNOW that there are no genetic gender differences in math ability, but apparently this radio host—along with much of the world—needed further evidence. After I argued this point to the host who was inter-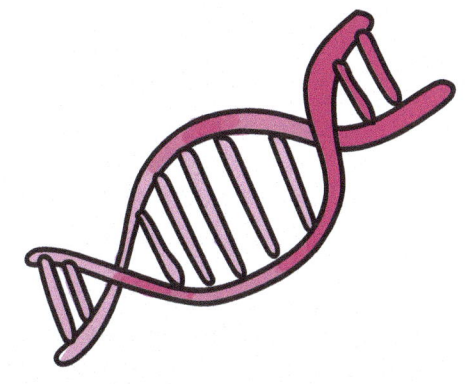
viewing me, he gazed off into the distance and then said—on-air—"Hmmm, right . . . but . . . I wonder what happens to kids' brains after puberty. Maybe there are gender differences in math ability at THAT point." I was SHOOK, but the interview ended, and I couldn't help but think to myself, Wow, all of this guy's listeners are now walking away thinking that there might be a genetic—AND GENDERED—component to math ability, despite ALL RESEARCH PROVING OTHERWISE. It made me so mad because *there is no math gene*. Let me say that again: THERE IS NO MATH GENE!

Despite the fact that it is exceedingly rare for anyone to have a brain that is so remarkable that it actually even remotely matters when it comes to being born with an exceptionality regarding math ability, the narrative that nature plays a significant role in mathematical intelligence has been around for literally ever. In 1879, French polymath Gustave Le Bon wrote that even in "the most intelligent races" there "are a large number of women whose brains are closer in size to those of gorillas than to the most developed male brains." He continued his insult with, "This inferiority is so obvious that no one can contest it for a moment; only its degree is worth discussion" (Pitman, 2023, p. 243). It would be nice to think that we've come a long way since then, but over a hundred years later, here we are trying to argue that men have better spatial abilities than women that make them more equipped to take on mathematics than their female counterparts (Bartlett & Camba, 2023), and it doesn't stop there. We still hear parents speak of a mythical math gene they feel their kids have—or haven't—inherited, despite research that there is *no such thing*. Given this pervasive, stubborn narrative, it's no wonder so many of us are scared of math: We've been told over and over again that we're either born "with it" or "without it" and that there's nothing we can do to change that. The result is that many of us feel an unrelenting lack of control and nagging paranoia at the thought that we might be one of the unlucky ones born at a deficit. Like, that's pretty terrifying if you ask me—downright scary. *Students aren't scared of math; they're scared of being genetically inferior.*

How many of your students' parents are under the assumption that math ability is genetic? Come up with some responses to keep in your back pocket the next time this comes up in conversation. You might cite the points I have made in this chapter, keep relevant links or articles on hand for interested parents, or even simply have a succinct "That's fake news" argument ready for when you need it. Helping your students' parents change their minds about math ability will ultimately help your students change their minds as well!

Math Makes You Miserable

Math class is tough!

On October 21, 1992, Mattel released a talking Barbie doll. When you pushed her little voice box thing, she would squawk, "Math class is tough." I'm not kidding—this was a real thing, and many young educators grew up WITH THIS DOLL because this was barely even 30 years ago, guys! The narrative that math is HARD is a pervasive one. Basically, every single Hollywoodized depiction of math paints it as a subject that only the nerdiest of the nerds can do (more on that later), and in our own schools, math gets a bad reputation for being the "hard" subject and the one students should avoid at all costs if they want to boost marks on their transcripts. I'm going to spare you my rant (for now) on how this bad rep is bolstered by the fact that kids are taught to fear hard things because they might lead to failure, and failure is bad and should be avoided at all costs (I'm rolling my eyes as I type this, but don't worry, we're going to deal with this pesky myth when we get to Chapter 3!). All that being said, math is positioned as being really, really hard—and that can be scary for most people because as a society (let's say it together): We hate hard things! *Students aren't scared of math; they're scared of being deeply uncomfortable.*

Can you think of any toys you had in your childhood that may have affected you in terms of your math ability? Maybe you specifically played with toys that helped you develop numeracy skills or spatial ability, or maybe you had a sibling who received very different messaging from the toys they played with? Think about it!

STRESS, ANXIETY, AND TRAUMA: OH MY!

In 2018, I was asked to do a talk about math anxiety for 100+ math professors at the Field's Institute, an international center for scientific research in mathematical sciences. It was one of the most awkward presentations I have *ever* given because, as my talk progressed, it became increasingly clear that my audience of (mostly white, cis male) math professors thought that math anxiety was total bullsh*t. They were legit rolling their eyes and snickering as I described what math anxiety looked and felt like, and when I polled the crowd, the consensus was that none of *their* students experienced anything of the sort! Ha! Yeah, right! What was even more shocking was that when I implored them to consider what areas of their lives they might have experienced anxiety in, they doubled down and insisted that anxiety was a made-up thing and that they didn't "believe in it." I'm not kidding!

In 2018, 9.4% of American children and teens were diagnosed with anxiety (Centers for Disease Control and Prevention, 2023). That number more than tripled to 31.9% in 2022 (National Institute of Mental Health, 2023) and is currently trending upward. I started my tutoring company in 2010, and I can tell you that at that time I literally never heard the word "anxiety" from my students. Sure, my teenage students would talk about how *stressed* they were all the time, but they wouldn't claim to have *anxiety* until years later.

Since anxiety is so often the product of underlying trauma, I think it's important for us to take a moment to really dig into what anxiety is and how it might present in our classrooms.

Math Anxiety

While **anxiety might feel like a buzzword in society right now, it's been around literally forever.** Simply put, the anticipation of potential threats in our immediate environment, which is what anxiety is all about, is actually just meant to keep us safe from danger. Back in the day, we really did need to consistently be on the lookout for danger. Think about it: At any point, we could legit be eaten by a tiger—like actually—so our brains developed this system to protect us from literal life-threatening situations. We really did need to be ready to fight that tiger— or run away from it—at any given moment!

The thing is, even though most of us aren't really at risk of being eaten by giant wildlife, our brains still sometimes think we are. Our brain's emergency hotlines flare up even when there's no real danger and the line starts ringing off the hook. This can cause those nervous, worried feelings we call anxiety. It's our brain trying to protect us, but it can get a little too overcautious, making us feel uneasy even when there's no real threat. While some level of anxiety is normal and can be motivating, excessive or prolonged anxiety can interfere with daily life. It involves both emotional and physical symptoms, such as increased heart rate, restlessness, and difficulty concentrating.

Now, similar to generalized anxiety, **math anxiety** is a heightened emotional response specifically related to the *perceived presence of math*. Take that in for a moment. Math anxiety doesn't just pop up when legit math is involved; all that is required to trigger math anxiety is the *perception* of math. For example, have you ever met someone who's like, "I'm SO bad with directions, don't even bother handing me that map!" or someone who's all, "Oh no, technology and I just don't get along" and gets into a frenzied panic when their laptop is frozen, subsequently smashing all the keys and having a near meltdown?! I'm telling you right now, in many cases, what lies beneath that panic is *math anxiety*! Now, if math anxiety can spike just at the mere hint of math, imagine how bad it can get when actual hardcore math is present—something that I'm sure you're faced with every day in your classroom!

> *Math anxiety doesn't just pop up when legit math is involved; all that is required to trigger math anxiety is the perception of math.*

Math anxiety might look like a student acting as if they're on edge in your class all the time, not just before a test. It might manifest as someone fidgeting, sweating, feeling sick, and even having trouble breathing. It might look like a student handing in a blank quiz paper even though a few days ago they knew the content, or it might even look like your student who regularly gets A's insisting that she sucks at math and feels like a total failure. It's incredibly important that your students know that anxiety has *nothing* to do with math ability. If your students have math anxiety, it doesn't mean they're bad at math, but it can impact their math performance because it decreases a cognitive resource called working memory, which means that when anxiety is present, there's less working memory to apply to the actual math task at hand (Boaler, 2014)!

Now that we have an understanding of what math anxiety looks like, let's talk about what is often lurking below the surface: math trauma.

Math Trauma

Trauma is a big word. Before I start talking about trauma as it relates to math and the classroom, I want to note that there is an entire field of study dedicated to trauma-informed education and ACEs (adverse childhood events), and that this is not that. I don't know about you, but when I first started hearing about trauma, I always assumed that, for something to be labeled traumatic, it had to be some giant dramatic thing . . . that is, until

I learned about *sneaky trauma* (also known as *microtrauma* or *little t trauma*). *Sneaky trauma* refers to smaller, often benign-seeming experiences, that make a person feel upset, insecure, or unworthy. We'll talk about this a little more in a sec, but first, I want to note that for the remainder of this book, when we talk about math trauma, we will be mostly talking about microtraumas that we as educators deal with in our classrooms every day. If at any point you feel that one of your students is experiencing something beyond the scope of your understanding or expertise, please reach out to your school leaders or appropriate mental health experts in your district.

> **If at any point you feel that one of your students is experiencing something beyond the scope of your understanding or expertise, please reach out to your school leaders or appropriate mental health experts in your district.**

It's really important to note that trauma refers to an emotional or psychological *response* to an event or series of events that are distressing or harmful and is often the catalyst for the development of chronic anxiety, or what is sometimes referred to as post-traumatic stress disorder (PTSD). When we talk about trauma, we are never talking about an event in and of itself, but instead about one's *experience* of that event. Two people can experience the exact same event, and one might experience it as sneaky trauma and the other might not!

I really like the way Gabor Maté talks about it because he explains why it is imperative for us to acknowledge and work on healing trauma:

> Trauma is a wound that leaves a scar or imprint in your nervous system or in your psyche and shows up in multiple ways that are not helpful to you later on . . . if you look at the nature of a wound, on the one hand if it's raw and open it really hurts, so if somebody touches that wound that you sustained a long time ago that you haven't healed yet, you'll react like you're being tormented all over again . . . on the other hand, wounds scar over and the scar tissue has certain features. It's very hard, it's rigid, it's not flexible, so people tend to be rigid when they're traumatized, and it also doesn't grow; trauma very often stops emotional growth and development. (Maté & Maté, 2022, p. 21)

We can't expect our students to take risks or exhibit vulnerability when they are protecting the tenderness of an open wound, nor can we help them grow and flourish when scar tissue is holding them rigidly in place.

MATH TRAUMA

Math trauma is an emotional or psychological response to an adverse event or series of adverse events that have to do with *math*. When I talk about math trauma in this book, I am talking about having a bad experience with math that sticks with you. It's not just finding math hard or boring or stressful; it's feeling scared, stressed, or even ashamed *because of* past experiences with math.

For many, traumatic experiences with math lead to debilitating long-lasting emotional and psychological consequences, which is why it is SO important that we gain a better understanding of how it occurs inside—and outside—of our math classrooms. As Jo Boaler so aptly puts it, *"Mathematics, more than any other subject, has the power to crush students' spirits, and many adults do not move on from mathematics experiences in school if they are negative"* (Boaler, 2016, p. xii). I'm sure we can all call to mind the many adults we know who likely have unpacked math trauma lurking beneath the panicked look that overcomes them when the check comes at dinner or their kind-of-bragging-but-low-key-embarrassed claims to not be able to balance a checkbook.

For many, traumatic experiences with math lead to debilitating long-lasting emotional and psychological consequences, which is why it is SO important that we gain a better understanding of how it occurs inside—and outside—of our math classrooms.

WHY MATH TRAUMA MIGHT DEVELOP

Before we dive deeper into how math trauma might present itself in our classrooms, I want to take a step back and really explain what sneaky trauma looks like and why math trauma might develop in the first place. Personally, I like to compare *sneaky trauma* to microaggressions. Hear me out: Microaggressions are small, subtle, or unintentional comments, actions, or behaviors that can hurt or offend someone.

For example, I was at a concert watching my absolutely favorite band, The Beaches (go listen to them RIGHT NOW! THEY ARE AMAZING! AND CANADIAN!). Anyway, The Beaches are a 4-piece, all-female band, and they are known for how incredibly energetic they are on stage. At the time of writing this, they have literally gone viral, and I bet by the time this book comes out, they'll have won 10 Grammys, and I'll be able to say, "Told ya so!" Well, this guy standing next to me at the show turns to me and goes, "I just can't get into them. There's just not as much energy when it's all girls on stage, you know?"

I. Was. Floored.

I would love to say that I said something witty and amazing in return, but I stood there truly shocked. I couldn't believe it. This guy was seriously insinuating that if there are no MEN on stage, the energy level scientifically DROPS? It was appalling and just straight up inaccurate, and what I really wanted to say was, "Oh yeah? What's your sample size? How many all-female bands have you ever seen perform?" I would bet you $100 his answer would have been: "One." This is a CLASSIC example of a microaggression. Microaggressions might seem minor, but they can make someone feel uncomfortable, insulted, or even excluded because of their race, gender, or other personal characteristics. When they take place consistently or repeatedly, they have a compounding effect and can be really damaging—*little t trauma* works the EXACT same way! We often don't label or unpack minor-seeming events that happen to us because they seem so benign, but the resulting experiences can be just as detrimental as those caused by *Big T trauma*. So, big or small, what types of events can cause math trauma in the first place? Let's take a closer look at three of the main areas of our students' lives where math trauma might happen: in the media, in the home, and in the classroom.

HOT TIP

Microaggressions are a mega cause of little t math trauma, and you can do something about it! If you ever spot any microaggressions in your classroom, call them out! Use such occasions as an opportunity to gently explain that jokey statements (e.g., anything emphasizing a stereotype) might seem light and funny to the person making the comment, but they actually reinforce damaging labels. These conversations are good opportunities for discussion and growth and go a long way in helping to cultivate a Math Therapy classroom!

Let's Blame Matt Damon

When I was in teacher's college, I did my first practicum at a local high school down the street, where most of my friends had gone when we were younger. On my first day, I walked into the math office and met with the head of the math department, who was this old dude who smoked cigarettes between classes and then PUT THE BUTTS IN HIS POCKET. He told us that that really pissed his wife off because when she threw his clothes in the washing machine you can imagine what kind of disaster ensued. (This has nothing to do with the story but I'm just trying to paint a picture here!) So I meet with the head of the math department, and the first thing he says to me is, "So, tell me. What is a pretty girl like you doing becoming a math teacher?"

And guys, this wasn't like, 1954. This was freaking 2008!!!

So where do whack ideas like this even COME from?!

Well, I have a question for you, and I want you to think long and hard: Have you EVER seen a movie in which the cheerleader character is good at math? Seriously. Have you? Okay, if by any chance you answered YES, I want you to literally put this book down and message me RIGHT NOW and tell me WHAT MOVIE because I have now asked this question to thousands of humans and not ONE person has said yes.

We all know the classic teen-movie plot: Hot, popular cheerleader lacks the smarts to get by in school. Her nerdy-but-lovable friend helps her pass math class, and in exchange, hot-popular cheerleader gives her a makeover so that the friend can get the guy she's been crushing on. From *Mean Girls* to *Clueless* to *The Big Bang Theory* to reality TV like *Love Is Blind* and *Too Hot to Handle*, this trope is literally everywhere! Just imagine being a girl who grew up watching those movies (which you may very well be!) and think about the message being sent: You can either be cool and popular OR you can be good at math . . . and a giant loser.

 IN THE MOMENT

Name one movie or TV show where the cheerleader character is also good at math. I dare you. My DMs are open . . . !

In graduate school, I studied the effects of media representations of math on teenage girls, the results of which culminated in my master's thesis, *Peace, Love & Pi: Imagining a World Where Paris Hilton Loves Math* (Vakharia, 2010). The results were shocking and heartbreaking. Every single one of the Grade 12 girls I interviewed told me that their social status in school was more important to them than being good at math, and more telling, that they all felt the pressure to choose one over the

other. Melissa, who was secretly scoring straight A's in her math class, told me that she felt as if she outwardly expressed her love of math she might risk having "no one to eat lunch with." She also ended up picking a career path that had nothing to do with math because as a girl who loves brand name designers, fashion, and celeb culture, she felt that there was a mismatch between "math's personality" and her own.

So what does this have to do with math trauma? Being told that the only way to belong in the world of math is to shed your literal identity CAN BE TRAUMATIZING! Just ask famed drag-queen-mathematician Kyne Santos (see

> **Being told that the only way to belong in the world of math is to shed your literal identity CAN BE TRAUMATIZING!**

Figure 1.2), who often talks about how they were made to feel as though they had to choose between EITHER being a drag queen OR a mathematician since no one would ever take them seriously if they chose to stick with both. Kyne is now making herstory with their book *Math in Drag*, but still gets told that they can't be taken seriously as a mathematician because of the way they look (which is laughable given they have a mathematics degree from one of Canada's top universities and over 1 million followers on TikTok who would agree that Kyne is *the only reason* they have finally learned to love math!).

Kyne is one of many who have been told that they must choose between being their authentic selves and being a mathematician (Gonzalez, 2023; Hottinger, 2017). Our students are being given this message by the media *every single day*. We hear it from the *Goodwill Huntings* of Hollywood that broadcast the message that to be a mathematician you need to be an awkward white dude with no social skills and prodigy-level mental math skills. We see it in the lack of diversity our kids are exposed to on their screens. Seriously, ask anyone if they've ever seen a Hollywood representation of Black people being good at math, and the only thing you'll hear anyone say is, "What about *Hidden Figures*?!" That movie was made in 2016 (and referencing the early 1960s!), and we're still hanging onto that TO THIS DAY because there has been NOTHING else; and we feel it in the social media trends that emerge almost monthly that tell girls they can either be "pretty" or good at math (see https://rb.gy/su72fi) or that reinforce the

FIGURE 1.2 KYNE SANTOS, AKA THE QUEEN OF #MATHTOK!

FIGURE 1.2 KYNE SANTOS, AKA THE QUEEN OF #MATHTOK!

SOURCE: Kyne Santos

stereotype that certain groups (like girls) are innately bad at math (see #girlmath [Ehsaei, n.d.]), while others (boys, certain racial groups) are innately good at math. Oh, and don't get me started on the constant bullying on social media platforms that serves to uphold the stereotype that if you're a girl and you ask questions you're an embarrassment to the entire FIELD of math (Goodyear, 2020).

This is not an exhaustive list, and I could truly go on about this forever (I mean, I DID write a whole thesis about it!), but my point is that these media representations of what it means to *be good at math* can be incredibly traumatizing for our students who are young people in the midst of identity formation, navigating who they are (and want to be) as human beings. Many of your students are being told that *who they are as people* is incongruent with the identity of someone *who is good at math*. Just think about that for a second. Imagine how much cognitive dissonance and stress that might cause a young person, and now imagine that young person being bombarded with that message multiple times a day. The traumatizing nature of math in the media is an example of little t trauma: We're so used to the way math is portrayed that we barely notice it, but the repeated, consistent nature of these microaggressions ultimately has a traumatizing effect on many.

HOT TIP

Students love a noncurricular task in math class! Challenge your students to find an example of math in the media and have them do an analysis of how math is being represented. Allow them their choice of media (advertising, social media, television shows, movies, even music) and ask them to present their findings to the class. This activity allows for rich discussion and debate about media representations of math and how they make your students see not only math but themselves as math learners!

The Apple Doesn't Fall Far

While most parents are well-meaning and wouldn't dream of intentionally traumatizing their kids, a lot of math trauma stems from family dynamics. Sneaky trauma in the home can be caused by parents telling their child that they "inherited their bad math skills" or that "the apple doesn't fall far from the tree" when it comes to math. Even though we now know—without a doubt—that there is no math gene, well-meaning parents still say stuff like this to their kids all the time; I hear it at least once a week from my students' parents! It is super traumatizing to feel as though your own parents don't believe in you—or even worse, that they are convinced that you were born with an innate math deficiency!

Being compared to a super-smart sibling can also be traumatizing, as can being yelled at by a family member for not understanding a math concept or for not getting a good grade on a math test. I love my parents, but I can tell you right now that I FOR SURE have math trauma that stems from my parents insisting that scoring 95% on a math test wasn't good enough. They would always say something like, "What happened to the other 5%?" and it sometimes made me feel like I wasn't smart enough and never would be, no matter how hard I tried. TMI, but this quest for perfection still plagues me TO THIS DAY—as I'm sure it does for many of you reading this.

In the Classroom

Math trauma often develops as a result of negative experience in our schools. These experiences might include things like public embarrassment, the pressure to perform, being singled out or labeled, a lack of student support, and classroom practices that can do more harm than good.

This is probably the scariest section to read because you might be like, "OMG, PLEASE tell me I'm not legit causing math trauma IN MY OWN CLASSROOM?!" But before you freak out: Remember, knowledge is power! Once you know what might lead to math trauma, you have the power to prevent it! Also, kids bring math trauma *into* your class *from* other classrooms, and it's important to know how and why that might happen so that you can help them heal from it. (*Spoiler*: We're going to deal with that when we get to the second step of Math Therapy, so you're already on the right track!)

Public Embarrassment

I have often heard people say something like, "Students don't hate math. They hate feeling embarrassed, ashamed, and defeated *by* math." If you ask any student what they hate about math class, one of the first things they will likely tell you is that they hate being called on when they don't have their hand raised. This is, more often than not, the result of a traumatic experience with public embarrassment. I remember all too well the feeling of sitting in math class SO stressed out that my teacher was going to call on me when I simply didn't understand the material being taught. I would sit in class with a knot in my stomach for the entire 60 minutes, PRAYING I wouldn't get picked to answer a question. When it inevitably did happen, my face would turn red, I would stammer that I didn't know the answer, and then I would sit there ashamed and embarrassed as my teacher made it clear that they were disappointed in me. Not a good feeling. Couple that with the experiences many of my students have had in similar situations where not only were they embarrassed in front of their teacher, but in front of classmates who would join in with a chorus of giggles. Totally traumatizing, especially for a young person!

 **IN THE MOMENT**

Can you think of a time where you might have felt embarrassed, publicly? How did you feel in that moment, and did it make you want to stop doing whatever it was that led to that embarrassment in the first place? If not, how did you find the strength to persevere? Keep this in mind around your students and use this as inspirational fuel to relate to them and to pep talk them when they need it!

Pressure to Perform

In 2023, I gave a talk to high school students at a "prestigious" private school about embracing failure in math class. I thought I was totally crushing it, channeling my inner self-help guru, being all like, "MATH IS ALL ABOUT MAKING MISTAKES, FAILURE IS AN OBSTACLE NOT AN ABSOLUTE, BLAH BLAH BLAH."

Imagine my surprise when, after what I considered to be a ROUSING speech, a group of students approached me and said, "Your talk was BS."

I looked around to make sure they were talking to ME, and when I realized they were, I was like, "Ummm, pardon me?"

"Your talk was total BS," they continued. "You say that we should embrace failure, but how can we when the whole point of high school math is to get into college, and we can't get into our top choice college without at least 90% in Algebra 1?"

Well, they had me there. I was stumped. They were getting mixed messages. Here, their teachers and admin were telling them that they needed to embrace risk-taking and failure and that nothing bad would happen to them as a result, but if you think about it, they were basically being lied to. Outside their classroom walls, they were being told that if they didn't adequately perform, they would be out of their top-choice school—something they had been told to work toward for their entire academic career thus far!

Being lied to? Misled? Being told one thing and then gaslit to believe another? Math trauma, math trauma, and MORE MATH TRAUMA!

Labels

When I was in Grade 7, I was sent to a school with a special program for "gifted" students. I had no idea why, nor did I know what being labeled "gifted" really meant. At my new school, all that label appeared to mean is that those of us in the program were weirdos. The kids in the other Grade 7 classes didn't  really talk to us, and we were treated like we should be smarter than them, always. Truthfully, I have no idea how my middle school experience would have been different had I been in any other Grade 7 classroom, but I do know that after that I never felt like I was allowed to "not get it" when it came to school. I was gifted. I was supposed to be smart. I wasn't supposed to struggle. That sentiment would follow me throughout high school, causing me to feel like a total failure when I started slipping in math class when Grade 9 rolled around.

Now, you might wonder what might be so bad about a label that would make a child feel "smart," right? I mean, isn't that better than labeling a kid as "slow" or "weak" or "behind?" The truth is that all labels can be harmful, even those that sound positive on the surface. Over the past few years, the gifted label has come under scrutiny (Mindshift, 2017) for several reasons, including the fact that it causes many children to feel like I did: like they are less-than if they ever struggle academically and

> **The truth is that all labels can be harmful, even those that sound positive on the surface.**

that their mathematical know-how is supposed to be a result of nature, not nurture. It can also be isolating for many students and can cause them to identify their self-worth solely with academic success, which can be incredibly unhealthy. However, just as with most things, there are pros and cons to not just the gifted label but *any* label meant to help educators identify how a child might benefit from different approaches to teaching (Lambert, 2024). In many cases, labels are useful because they help us categorize and understand, and doing away with them entirely isn't as necessary as understanding that labels carry baggage and that we can help our students unpack that baggage with compassion and grace.

Given the influx of assessments, IEPs, learning differences, and mental health diagnoses that now permeates our classrooms, it is important for us as educators to help our students understand that they are *more than* the labels that they have been given! Labels are useful when they help us adjust our classroom practices or teaching methods to meet a student's needs but harmful and traumatizing when they are used to dismiss or categorize students in ways that limit the way in which they are understood as complex, ever-changing individuals.

Lack of Support

You might be thinking: The whole reason we have labels is so students can receive the extra support they may need. But what happens when that support simply isn't available? Sadly, we all know that with increasing class sizes and cuts to education budgets, that is what we as educators face every single day. As hard as we try to honor all of our students, there is usually only one of us there to serve 30 to 40 students with diverse needs in each one of our classrooms. A lack of teacher support ultimately leads to a lack of student support, and labeling our students without bolstering those labels with the support needed can lead to trauma, as it sends the message to students that they are difficult, hard to handle, or even annoying, and that there is nothing we can do to help them.

Let me be loud and clear: This is not the fault of us as educators. This results from the failure of the system we work in. Nevertheless, acknowledging this downfall is so important in helping our students heal. Labels aside, even our unlabeled students end up feeling those feelings of inadequacy and defeat if they are unable to feel supported when they are struggling. This might include students with learning gaps, those who are unable to afford support outside the classroom, and even those who face barriers that prevent them from getting homework done because they are needed to care for siblings at home or must work after school. Helping these students feel supported by acknowledging their circumstances and showing empathy can go a long way in preventing math trauma and healing math trauma that has already formed.

Classroom Practices

Okay, I promise I'm not going to tell you that everything you're doing in your classroom is horrible and traumatizing, I promise—PROMISE!

What I will say instead is that many of the classroom practices we as teachers grew up with may have caused math trauma . . . for US! Math education is currently undergoing a revolution, and part of that revolution includes reexamining the classroom practices we have become so accustomed to and investigating how they may be part of the reason *so many adults* today HAVE math trauma!

Let's take timed tests, for example. Current research shows that the use of speed to measure understanding actually causes students so much anxiety that they underperform relative to their actual mathematical knowledge (Boaler, 2014). Not only do timed tests not accurately measure how much a student actually knows, but they lead to high levels of anxiety and negative experiences with math that ultimately cause math trauma! Now, there are many classroom practices just like this that we educators have been using for years because they've become the fabric of the systems in place. These practices include the way we group and assess students, the method used to teach certain content, how we create rubrics, and much more. I'm going to save this discussion for Chapter 4 when we talk about how to unpack and heal math trauma because we're going to go over ways in which you can adjust some of these practices so that they satisfy the requirements laid out by your district, while simultaneously taking into account their potential to cause math trauma.

WHAT MATH TRAUMA MIGHT LOOK LIKE IN THE CLASSROOM

I'm going to end this chapter with a snapshot of what math trauma might look like in our classrooms based on the four classic trauma responses: fight, flight, freeze, and fawn. The four trauma responses evolved as survival strategies in the face of danger and developed over thousands of years as adaptive behaviors to help early humans respond to threats and ensure survival. As I mentioned earlier in this chapter, while these responses were beneficial for survival in dangerous situations, today these instincts can still kick in when we face stress or perceive danger, but they might not always be as helpful in modern situations as they were back in the day!

Can you spot any of these four trauma responses in your classroom? Bonus points if you go through your class list and give some thought to each of your students and what behaviors they typically exhibit that may be indicative of one of the four trauma responses. Save this list for later, as we'll want to revisit it in future discussions!

FIGHT
- irritability
- anger
- aggression
- "moving toward"

FLIGHT
- panic
- anxiety
- perfectionism
- "moving away" (avoiding!)

FREEZE
- spacing out
- feeling stuck
- dissociation
- depression/shame

FAWN
- people-pleasing
- avoiding conflict
- difficulty saying no
- prioritizing others

Fight

Our *fight* response developed as an evolutionary protection so that we could defend against danger. Remember that tiger we talked about earlier or other equally terrifying threats we were always dodging. We needed to be prepared to legit grab our spears and fight back! In our classrooms, a student who has experienced past math trauma might go into fight mode when they sense danger. Sure, a math test doesn't seem quite as threatening as a tiger, but our amygdala doesn't exactly know the difference—the same emergency hotline system gets activated regardless of the danger! This fight response might manifest as frustration or defensiveness, competitiveness, anger, or even hyperactivity and overachieving.

Flight

Let's face it, not everyone wants to tackle a tiger with a spear. *Flight* is the trauma response that instead tells us to RUN FOR OUR LIVES instead of facing a scary situation head on. You can see how both fight and flight would be potentially useful . . . depending on the size of the tiger (or math test)! In our classrooms, we tend to see flight A LOT. Flight commonly manifests in student behaviors that might look like avoidance, disinterest, and distraction.

Freeze

Sometimes when in danger, staying really still could make someone harder to notice, like playing dead might work to avoid being seen by . . . you guessed it . . . a giant tiger! Think of it as being as quiet and motionless as possible to avoid being a target. In the classroom, a student might feel completely paralyzed when confronted with a challenging problem. We see this in students who suddenly draw a blank when faced with their test paper—picture those who sit quietly, unable to attempt the question, essentially *freezing* in the face of the challenge (Kubala & Lebow, 2022).

Fawn

Finally, our *fawn* response developed as a way to appease or even befriend a potential threat to reduce the risk of harm. Back to that tiger: Instead of fleeing the scene or battling this giant mammal, perhaps you might choose to pet the tiger . . . make soothing noises . . . or become its bff to avoid getting eaten? You do you! In our classrooms, this might show up as a student seeking constant approval, overapologizing for not understanding a concept, agreeing with what the teacher says and claiming to understand even when they don't, and an inability to assert needs even when asking for an accommodation might actually help them feel better about the math they're learning.

Which of these 4 F's is your style? I'm totally a flight girl when I'm dealing with relationship drama, but a fawn girl when I'm dealing with online haters . . . what about you?

HOT TIP

Keep an eye out for the 4 F's in your classroom! Seeing your students' behaviors through the lens of the 4 F's can help you view them through a lens of empathy and compassion rather than the frustration that it's totally normal to feel when we see our students struggling and don't know how to help. Which of the 4 F's can you spot in your classroom? What do they look like? Which students exhibit them and when? Go through your class list and make notes; these will give you valuable insight as you go through Math Therapy with your classroom!

BEFORE WE MOVE ON

Not so fast! I know you're excited to get to the next chapter, but first—let's celebrate our progress and let it all sink in!

Treat Yourself

I am a big believer in celebrating every little bit of progress we make in this life because, hey, life is short, but also because we are often so fixated on the final destination that we NEVER take the time to celebrate the journey! At the end of every single chapter, I'm going to give you a teeny, little something sweet to do so that you can *feel good* about what you've learned. I want you to take these moments to use what you've learned in this chapter to *treat yourself*, okay? Don't skip this; it's the best part!

This end-of-chapter treat is all about showing yourself grace. Stress, anxiety, and trauma are parts of *most* of our lives, so don't be hard on yourself when they pop up. Instead, realize that they are just a part of you, like everything else.

So: I want you to think about one amazing character trait that you have that has gotten you through a particularly stressful time in your life. Picture that stressful time and then hone in on the amazing, magical trait that got you through it. Can you see it? Can you *name* it? Seriously, speak its name into the ether. Now, I want you to take your magical trait on a hot date! Time to treat the trait that's gotten you through a tough time by doing something fun. Take it out to dinner, give it a rom-com movie night, throw on twinsie face masks—do something nice to treat your trait (aka one part of YOURSELF that ROCKS) before moving on to the next chapter. You deserve it!

ASK YOURSELF

This is your chance to process by reflecting on what you've learned in this chapter.

1. Did this chapter change your perspective on why so many people are scared of math? Why or why not?

2. Are the ideas presented in this chapter different from your current educational philosophy? If so, how?

3. Can you think of examples from your classroom that illustrate the difference between stress, anxiety, and trauma?

4. What surprised you most about the causes of math trauma listed in this chapter? Which cause did you find most surprising?

5. Can you think of an example of how math trauma presents itself in your classroom? Describe it in detail!

6. Will any of the information learned in this chapter change your classroom practice, even just a little bit? What and how?

☮♡π

CHAPTER 2

WHAT IS MATH THERAPY?

In this chapter, we get to:

1. Learn the 5 M's of Math Therapy

2. Discover why I created Math Therapy in the first place

3. Understand the interdisciplinary framework that upholds the Math Therapy process

4. Get familiar with what Math Therapy is . . . and isn't

5. Learn why anyone (yes, including you!) can be a great math therapist

Okay, that last chapter may have been a little scary, am I right?! I'm not psychic, but as I'm typing these words, I'm imagining some of you huddled under a blanket being like, "OMG, I just realized that I miiiiight have more trauma lurking in the dusty corners of my mind than I thought . . ." (Or maybe that's just me!?) The good news is, now that we're

SOURCE: Racheal McCaig

all on the same page about math trauma, we can move on, *together*, in the same positive direction. You know what else is good news? Even if you're like, "Ummm, nope, no math trauma here," you're going to learn how to PREVENT math trauma from happening in your classroom in the first place, which is of equal, if not greater, value. So wherever you are—I'm glad you're here!

THE 5 M'S OF MATH THERAPY

THE 5 M'S OF
MATH THERAPY

STEP 1: Mythbust

STEP 2: Moderate

STEP 3: Motivate

STEP 4: Makeover

STEP 5: Measure

Okay, look, I don't want you to accuse me of being one of those authors that makes you wait until the final page of the book to finally get to the point, so I'm just going to tell you RIGHT NOW that Math Therapy is a 5-step process that you're going to use to help your students heal their math trauma and build a better relationship with math. There—THE SURPRISE HAS BEEN REVEALED—now you can't say that you didn't know! A Math Therapy classroom is a learning environment where students are consistently engaged in building better relationships with math—and, as a result, with themselves!

The 5 steps of Math Therapy are meant to be carried out sequentially at first, and then repeated consistently as needed. After the initial 5-step process has been implemented once, you can decide which steps need to be revisited or reinforced—don't worry, we'll get into more detail later in this chapter! For now, let's go through this step by step *briefly* so you know what you're getting into. Then in the chapters that follow, we will be doing a juicy deep dive into what each step entails and how to carry it out within your classroom—AND I'm going to be giving you tons of strategies, tools, resources, and templates to use to make it super easy! Ready? Let's do this!

Step 1: *Mythbust*

You will use the first step of Math Therapy to *Mythbust* **your students' misconceptions** about what it means to be a "math person." (*Spoiler*: There is no such thing—but you already knew that!) In this step, you will use loads of class-room strategies to convince everyone in the room that we are all capable of doing math in a meaningful and rigorous way.

Step 2: *Moderate*

Once you've convinced your students that it is 100% possible for them to do math, you're going to move on to Step 2, which is all about helping students *Moderate* **their math trauma!** Moderating math trauma is all about making space to see what comes up and reflecting on it with a fresh perspective. Even if students don't necessarily resonate with the idea of math trauma, this step will help them unpack their relation-ship with math, and everyone has something to gain from that!

Here, you will introduce your students to the concept of math trauma, create an environment in which they feel comfortable sharing their own past experiences, validate them for the complicated feelings that often accompany those experiences, and work together as a collective to create a space in which retraumatizing students can be minimized—even eliminated altogether!

Step 3: *Motivate*

At Step 3, your students should know and believe that their brains are capable of doing math. They understand that their complicated emotions around math are a result of past math traumas, that they're not alone in feeling this way, and that those experiences and emotions don't define them (as people or as doers of math) or what they're capable of. Now, you have to **Motivate your students to keep going!** In this step, you're going to raise the stakes for all of your students and tap into a variety of tools and strategies to convince them to CARE about their relationship with math!

Step 4: *Makeover*

Next, you're going to help your students *Makeover* their math stories! Now, you might think that this part is only important if a student has a less-than-ideal relationship with math, but that's totally not the case. Every single one of our students can benefit from reflecting on the intricacies of their math story in so many ways. Not only does articulating one's positive mathography help to reinforce a positive math identity, but if you take it to the next level and encourage students to share their math stories with one another, this can be such a great way to model the diverse ways in which positive relationships with math might play out: Representation is important—you can't be what you can't see!

As we have explored in the Mathography assignment at the beginning of the book, we know we each have a story that we tell ourselves every day. In this step, you will not only teach your students how to give their math stories a makeover, but you will show them how to bring those new stories to life!

Step 5: *Measure*

It's almost time to celebrate—but first, you've got to show those kids that *this is working*! This step is all about helping students **Measure their progress** in a real and tangible way. In this final step, you will use a ton of action-packed strategies to allow your students to come to their own *aha!* moments in a variety of unique and meaningful ways!

* * *

You now have a full picture of what the process looks like and you're ready to dig in and get started . . . right!? Before we do, remember, as educators (and as human beings who have lived on this earth), we all know that true mastery comes from repetition and consistency. Math Therapy isn't something that just happens once, but something that must be practiced *every day*. That's why, throughout this book, we'll be talking about how to weave Math Therapy into every aspect in your classroom culture so that your classroom becomes—POOF—a Math Therapy classroom!

 **IN THE MOMENT**

Take an inventory of your students. Who do you think might benefit most from each step and why? Thinking about this before you embark on Math Therapy can give you ideas on how you might focus each of the steps as you move through the chapters and learn more about what each M entails.

BUT WHAT DO THE 5 STEPS ACTUALLY LOOK LIKE IRL?

I totally get that without an irl example, it's hard to understand how these 5 steps actually work, so I'm going to tell you a story about how I carried them out with a guest on my podcast! Now, I know that most of you are reading this book because you want to do Math Therapy in your classroom with students, but trust me, this one-on-one example gets straight to the heart of the 5 steps in a quick and direct way! In this case, because I was limited to 60 minutes during a podcast everything had to happen quickly. This is NOT going to be the case for most of you. But this one-on-one example will give you the big picture. Ready? Meet: Tiana!

Step 1: *Mythbust*

During the first season of my podcast, I interviewed a woman in her mid-20s that we're going to call Tiana. I had met her at a bar and told her I had a podcast called *Math Therapy*. You can guess exACTly what happen next: She shuddered, convulsed, practically spit out her drink, and then said something along the lines of, "EW I HATE MATH I COULD NEVER DO IT I DON'T UNDERSTAND HOW YOU CAN BE GOOD AT BOTH MUSIC AND MATH THAT'S SO WEIRD EW EW EW." I calmly explained that I TOTALLY knew how she felt, and in fact, her reaction was not unusual, and actually, one of the reasons I started the podcast was to help people like her see that her dramatically adverse reaction was likely a result of underlying math trauma, and actually, that she was likely more capable of math than she thought. She was adamant that no way, no how would she ever be capable of math and that there was just something intrinsically wrong with her. I started to get excited (who doesn't LOVE when the *perfect* Math Therapy candidate walks up to them AT A BAR!) and told her that I, in fact, had once felt the same way and proceeded to regale her with my tale of failing Grade 11 math TWICE and my eventual metamorphosis into a math lover who had scored a whopping 100% in first-year university calculus.

She had never heard a story like that before, she told me, and she was unsure what exactly to do with this idea that people could transform when it came to not only math ability, but the subsequent enjoyment of math that might follow. You can imagine HOW excited I was: Tiana could clearly see that not only had I apparently transformed from someone who FAILED math into someone who was GREAT at it, but into someone who loved it SO much that here I was accosting this random person at a bar at 1:00 in the morning to tell her ALL about it. I asked her if she wanted to be a guest on my podcast about Math Therapy, and to my surprise, she said yes!

In Tiana's case, it turned out that Step 1 had happened without much intentional effort on my part. Instead of directly preaching the intricacies of growth mindset and neuroplasticity and how our brains worked, I had relayed the entire message by sharing my own story.

Step 1, complete!

Step 2: *Moderate*

Once the interview started, I asked Tiana if she could remember why it was that she thought she wasn't a "math person." Her initial response will sound familiar to you because it reflects what most people will say when asked this question: "I wasn't good at it." I probed further, asking her what that meant to her exactly. She came back with a few tentative, but telling, explanations:

"I made a lot of mistakes."

"Even though I tried, I didn't get good grades."

"My mom was super nice about it, but yeah, she always just told me not to stress out about math because there were lots of things that I was better at."

I thanked Tiana for being so vulnerable and digging deep and told her that what she shared sounded really hard, stressful, and kind of traumatizing. I shared that, even though it might seem benign and that I'm sure her mom was well-intentioned, that repeatedly telling someone that they shouldn't worry about math because they have "other strengths" inadvertently sends the message that there's no *point* in them trying to get better at math because it's likely *not possible*. She had never thought about it that way before and agreed that actually, she may have gotten that idea along the way without even realizing it!

We chatted a bit more and then she said something that piqued my interest: "Oh, I forgot to tell you, something really exciting happened last year! I was FINALLY diagnosed with ADHD at the age of 27!" Tiana laughed, continuing with, "Life is SO much better with Ritalin, lol."

I pondered out loud, "What do you think would have happened if you had been diagnosed earlier and had been able to do math on Ritalin?"

"Ha," she responded, "it probably would have been a LOT easier!"

I was thrilled. Why? Because I was about to blow Tiana's mind.

I gently explained that it can be really traumatizing to think that there's something wrong with you, only to be told years later that you actually have a legitimate, diagnosable condition, which wasn't just overlooked and ignored, but which comes with a set of resources that you didn't have access to until MONTHS ago. Her eyes widened as she realized the magnitude of this revelation. An undiagnosed learning difference certainly isn't a component of every case of math trauma, but it can be a contributing factor. We had completed Step 2 with a BANG. Tiana understood that both her mother's well-intentioned words, coupled with her late-stage ADHD diagnosis, may have contributed to underlying math trauma that prevented her from feeling capable of doing math and of taking the steps she needed to get there. She felt validated for those feelings, and she acknowledged that that trauma was not fixed nor did it define her. It was simply an obstacle to be unpacked and understood along her math journey, which as it turned out was just beginning.

It was time for Step 3.

Step 3: *Motivate*

I lucked out with Tiana because when she had initially agreed to take up my offer of FREE Math Therapy, courtesy of the podcast, I had a suspicion that there must be an underlying reason why she said yes. So I asked her, point blank, why she had agreed to being on the podcast, given that she was far beyond having to take math ever again in her life at this point.

Tiana started by explaining that she felt that her insecurities around math were holding her back in her career. Tiana worked in radio, and there was a job she had been wanting to apply for that she just couldn't work up the courage to go for: It was an assistant director role that would require her to have a quick grasp of clock math. So for example, if a host was to be on air at 10:00 a.m., at 9:59:50 a.m., Tiana would need to prep the host by saying, "And you're on in 10 . . . 9 . . . 8 . . ." and so on. She didn't feel confident that she could do that. I, on the other hand, was confident that she *absolutely* could.

She continued by mentioning that she felt that her lack of confidence around math was part of the reason she had gone into credit card debt. She proceeded to tell me a heartbreaking story about how months earlier, she had seen a sale sign on top of a pile of coats at Aritzia. She quickly calculated what she thought the sale price was and then brought a pile of coats to the cash register. The thing is, she TOTALLY calculated the sale price incorrectly, and the coats were WAY more expensive than she thought, but she had so much shame around admitting that she had done the math wrong that she bought the coats anyway. What proceeded was a shame spiral, which included shoving the coats in a corner of her apartment and refusing to wear them because of what they symbolized, followed by months of credit card debt.

Step 3 complete—she was motivated!

Step 4: *Makeover*

Tiana said something really interesting in Step 2—did you notice? Here it is again:

"Even though I tried, I didn't get good grades."

BAM. There it is. It might seem like a casual comment to drop, but it's actually a golden nugget for us educators! That comment is a sneak peek into the math story she's been reciting in her head ALL these years. That and the fact that she's already blatantly said, "I'm not a math person"—that's part of her story too. But I have to be honest; the comment about her trying is much more interesting because if you think about it, if she honestly believes she has tried every single avenue on the planet and still can't do math, it makes total sense that she thinks she's not a "math person." The thing is though, I know she hasn't. And I'm about to prove it to her.

V: Tiana, what did you mean when you said, "Even though you tried"?

T: I mean I tried everything! I did my homework, I studied for tests—none of it helped!

V: Out of curiosity, when you did your homework, did you check the answers to make sure they were right?

T: No.

V: So how do you know you weren't, like, practicing the wrong thing?

T: I guess I don't . . .

V: Did you ever get a tutor or anything?

T: No.

V: And you already told me that you weren't diagnosed with ADHD yet, so you probably didn't have access to any resources like medication or coaching?

T: No.

At this point, Tiana is starting to see my point before I've even made it, which is that, although she struggled and that struggle may have made her feel like she was at the end of her rope, she actually had a TON of rope left—but no one to help her access it! It is generally very empowering for students to know that they are NOT trying their hardest, and I mean this in the nicest possible way. I'm not calling Tiana lazy or unimaginative or anything pejorative. What I told Tiana is that while it's impressive that she tried her hardest at the time, that due to no fault of her own, she wasn't made aware of all of the other avenues available to her in terms of her building a better relationship with math. I also told her that this happens to SO many people, she's not alone, and that I myself had a tutor, went to an alternative school, and studied for 5 hours a day in my last year of high school to finally feel really good about math. It's not like I changed my mindset and automatically turned into, say, Beth Harmon from *The Queen's Gambit* or something. I had to work really, really hard, and that was just my personal path. Some people don't have to put in as many hours, some people need to put in more hours, some people have a teacher who teaches in a way that jives with them, some don't, and so on. The point is that it was time to rewrite Tiana's narrative:

> Before: *I tried everything, and I'm hopeless at math.*
>
> After: *I tried some things, and thought I was hopeless at math, but if I had access to more resources, who knows how much better at math I could have gotten!*

Do you see how much more empowering that second statement is? It is rife with potential for change, curiosity, openness to the unknown—and *hope*.

By the way, Tiana and I rewrote her narrative together, and we decided that, due to the fact that we only had 1 total hour together on the podcast, that we would work on the core belief rather than on her entire math story (you'll get to choose what you want to do in your own classroom—there are options)! In this step, we tried a few different phrases until we found one that seemed true to her. Notice that the facts don't really change, but the sentiments around them do. Now that we had her new mini-story, she was ready for the final step.

Step 5: *Measure*

With 10 minutes left to go in our interview, I needed to show Tiana that the hard work she had put in was already working. I reminded her that one of the most exciting facts about the brain is that it is literally always growing and that her brain at this point of our interview was literally *different* than the brain she had when she walked into the studio. I reminded her that any math she elected to do was optional but that since in Step 3 she had told me that one of her motivations of being on my podcast was to help her avoid future credit card debt, and since her last foray into said credit card debt had been due to a sale price miscalculation, would she be open to me attempting to teach her how to accurately calculate a sale price on the fly? The first sign that Math Therapy was working was that she didn't even hesitate before saying yes! And off we went. Here is the example I started with:

A dress originally costs \$50. It's 30% off. How much does the dress cost on sale?

Step 1: Subtract the sales % off from 100. In this case, it's 30% off, so:

$$100 - 30 = 70$$

Step 2: Turn that number into a decimal by dividing by 100.

$$\frac{70}{100} = 0.7$$

Step 3: Multiply the resulting decimal by the original price. In this case, the original price of the dress was \$50, so:

$$\$50 \times 0.7 = \$35$$

The final sale price of the dress is \$35.

I taught Tiana three total steps and made sure she understood why we were doing each one. My goal in this step wasn't to get Tiana to understand every mathematical concept she had never grasped before, but to show her that she *was* capable of doing something she previously felt incapable of doing—that was my only goal here. We ran through the calculation five times, each time using different numbers, until Tiana felt fully confident. At the end, she asked if she could snap a photo of her work and save it on her phone so that she had an example to look at next time she had to calculate a sale price. I suggested she keep an eye out on the subway, in her coffee shop, and in the grocery store and that she calculate one sale price a day to get the process into her muscle memory. She eagerly agreed, and after an hour together, Tiana walked out of my studio thrilled that she had learned to do something she previously believed was beyond reach.

* * *

A few weeks later, my phone vibrated with a notification; Tiana had sent me a DM (yes, via TikTok, IF YOU MUST KNOW) to tell me that she had mustered up the courage to apply for the assistant director position at her radio station and had *gotten the job*! She had also finally dug her way out of credit card debt and was feeling confident about her ability to stay on top of her finances. Math Therapy had served its purpose: It had helped Tiana rebuild her relationship with math to a point where she felt confident enough to take the risks she perceived she needed to be happier, period.

I don't know whether Tiana continued exploring her relationship with math or not, but that part doesn't actually matter. Remember, the goal of Math Therapy isn't to skyrocket everyone into a STEM career or to make mathematicians of us all—it's to empower us to build a better relationship with math so that we can ultimately *feel good about ourselves and our life choices as a result*. And for Tiana, that's exactly what happened.

In the five chapters that follow, I'll be showing you examples of how to execute the 5 steps of Math Therapy in any setting—from a prison (really!) to a classroom, with real-life examples, templates, resources, tools, and strategies to allow you to implement each step RIGHT AWAY! But don't you want to know the story of how I came up with it first? I thought so. Grab a cup of tea and get ready to travel back to a simpler time with me (aka a time when we only wore masks on Halloween, Britney Spears and Justin Timberlake were in rom-com love, and "tiktok" was the sound that a clock made).

HOW DID I COME UP WITH MATH THERAPY ANYWAY?

Things are about to get juicy, and I'm so glad that you're along for the ride. I promise that I didn't just include this section to make myself look cool—I actually think it's really important to give you the context surrounding how Math Therapy came to be so that you can start YOUR Math Therapy journey from the same place that I did!

"And That's Why I Became a Journalist . . ."

A few years ago, I was asked to be on the *Global News* morning show to talk about math anxiety. Now in Canada, *Global News* is a national show, which means that it is shown to millions of viewers every morning. So there

I am, on the show, and as the segment starts, the host says something like, "So Vanessa Vakharia is here to talk to us about math anx—oh ugh, MATH. Who DOESN'T have anxiety around math? I HATED math when I was a kid—that's why I became a journalist . . . HA HA HA." Picture my face, guys. Not pretty. I was, like, trying not to look completely appalled by the fact that this guy was totally pushing the everyone-hates-math stereotype ON NATIONAL TELEVISION, but before I could get a word in, he went on: "You know, I'm so bad at math, I need to use my TOES to count to 20!" And then, he did the unthinkable. He proceeded to REMOVE HIS SHOES. AND COUNT TO 20. ON HIS FREAKING TOES. ON NATIONAL TELEVISION. (See Figure 2.1 for a photo that does NOT show bare feet.)

I. Was. So. Embarrassed. For HIM!

But also, I was kind of livid on behalf of all Canadians who had just watched that segment and had walked away with the idea that math is just some gross, horrible thing that some people just can't do, and that's why they become . . . journalists?! It was just another nail in the "math is the worst" coffin, and I was over it.

FIGURE 2.1 PHOTO OF ME AND LINDSEY DELUCE ON CANADA'S CTV'S *YOUR MORNING* (SHE'S A HUGE ADVOCATE FOR MAKING MATH MORE ACCESSIBLE)!

SOURCE: Racheal McCaig

NOTE: This is a *completely* different news segment than this story is about—for obvious reasons!

HOT TIP

Always choose your words carefully when you're talking about math ability in front of your students! Are you sending the message that you're stressed out about math or that you don't believe that *you* can ever be good at math? You might be unintentionally sending your students mixed messages, so be careful!

FIGURE 2.2   SABINA AND ME IN
PODCAST RECORDING MODE BACK
IN 2020!

The truth is that this wasn't the first time this type of thing had happened to me. In fact, because I do so much work with the media, I am often faced with journalists looking pained and totally freaked out when they discover that they're interviewing "The Math Guru" on their segment. I can SEE the math trauma in their faces, almost like they're scared I'm going to give them a timed test ON THE SPOT, and they'll be forced to relive all of their math fears irl, on TV. So this particular *Global TV* incident wasn't new, but it was so extreme that I walked away and said to my friend Sabina Wex (see Figure 2.2), who was doing PR for me at the time, "Wow, that guy needs *math therapy*."

 **IN THE MOMENT**

Have your students think of a time they have heard an adult talk about math. It might be a parent, teacher, or friend. What did that person say or do? Did it feel like that person had a good or bad relationship with math? Why? Did it change that student's mind about how *they* felt about math themselves? Why or why not?

Sabina, David Kochberg (my ex-boyfriend/now-bff/technical producer/the other half of my band Goodnight Sunrise—it's complicated), and I started the *Math Therapy* podcast in 2019 and never looked back. It started out as a fun way to give actual people Math Therapy irl on the pod. When we started it, most of the guests we had on the show were people who hated math or had math anxiety, just like that journalist who took off his shoes to count to 20 on national television! From students to screenwriters, drag queens to drug dealers, professors to PRISONERS—we've heard over 50 tales of complicated relationships with math on our podcast! As we began to realize how pervasive, insidious, and complex math trauma could be, *Math Therapy* quickly turned into more than a podcast and morphed into a formal framework that would change the lives of not only my own students, but the lives of other educators (just like you!) and THEIR own students!

The thing is, as Math Therapy was *formally* taking shape in 2019, I was starting to realize that I had been sort of doing Math Therapy . . . since I was in high school! Let me explain.

The More the Merrier

Have you ever noticed that some of the coolest things on the planet are a result of mashing up a bunch of things that might not traditionally go together? In my opinion, mashups are the literal lifeforce of innovation. Just think of all the amazing music mashups there are out there! Michael Jackson and Paul McCartney's collab on "The Girl Is Mine" is legit legendary, and OMG, don't even get me started on the new version of "Hold Me Closer" that Elton John released featuring BRITNEY SPEARS! Iconic! But mashups aren't just relegated to music; they're the source of innovation in so many industries! Think of the amazing food mashups out there—like the Cronut (croissant + donut) or sushirrito (sushi + burrito) or, I mean, literally ANY combo of sweet and salty—is there anything more delicious?! What about the mashups the tech industry has reinvented itself with, hello! GPS (Global Positioning System) technology emerged from combining principles from the fields of astronomy, physics, and computer science, and most of us now use that every single day on our smartphones—which, by the way, are *also* a result of an epic mashup! In fact, the first commercial camera phone was the Kyocera Visual Phone VP-210, released in Japan in May 1999 (Dobrow, 2022), essentially a camera MASHED UP with a cellular phone: Look at how far THAT went!

HOT TIP

Don't be afraid to bring your whole self into math class by mashing up math with the things you love! Are you a sports fan? Use the same motivating techniques coaches use to motivate their players! Love music? I have heard so many stories from teachers about helping students familiarize themselves with math concepts by having them write formulas and explanations right into the lyrics of their fave songs!

Mashing up fields of study is what interdisciplinary work is all about, and I have never understood why education as a field seems so resistant to it while other fields literally *thrive* on it. To combat this siloed approach to education, I wrote a thesis for my master's program to include perspectives from marketing, gender studies, and math education to explore the reasons why so many female students get the idea that they're less capable of math than their male counterparts.

A Focus Group . . . for Math?

Here's one of my marketing-based arguments: Companies use focus groups to figure out what will appeal to the demographic they're trying to sell to. If a company wants to figure out how to sell teenage girls eyeliner, for example, they get groups of teens in a room and show them different products to ask what they think about the product, the packaging, and so on—or they show them different ad copy about the eyeliner to see what resonates with them.

Why don't schools do the same thing? We don't treat our students as an audience that we have to appeal to; we just take them for granted. We don't *care* if they like the "product" we're selling (in this case, math curriculum). We don't ask them what would make math more appealing, what kind of copy they'd like to see on their textbooks, what type of delivery would convince them to "buy" in. And as a result, the majority of our students are disengaged and feel as though math is something that is forced upon them, not something they would *choose* to consume if given the choice. But why? Education policymakers have the same access to marketing techniques (like the use of focus groups) that companies like Coca-Cola or Maybelline have. So why don't we make use of that knowledge and of those strategies to make OUR product more valuable to OUR target demographic?

> *As a lifelong teacher, it has—and always will be—my mission to turn meaningful connections into a rich and rewarding way to help students build an equally rich and rewarding relationship with math.*

A diverse mix of voices, perspectives, and bodies of knowledge leads to better discussions, decisions, and outcomes for everyone, period (Gomez & Bernet, 2019; Phillips, 2014). That's why as a lifelong learner I have always found it important to find the meaningful connections between everything I have learned in this lifetime. As a lifelong teacher, it has—and always will be—my mission to turn meaningful connections into a rich and rewarding way to help students build an equally rich and rewarding relationship with math.

The thing is, because of my own complicated relationship with math growing up, I instinctively always knew that learning math was more about the context in which it is taught than the content that is ultimately delivered. By that, I mean that I had experienced it myself: content couldn't reach certain students (me included) unless the context in which it was taught changed. For me, that meant going to a different school where my teachers approached math education from a place of belief that *there was no such thing as a math person*, instead of a place of *some students have it, and some are meant to flip burgers* (that is a direct quote from Miguel, one of my Math Therapy guests, by the way. He was told the only thing in his future was fast food, and he now does statistical analysis for the Ontario government (Vakharia, 2020). My point is, at my old school no amount of content was ever going to shift the way I felt about math—or myself—until the context in which it was taught changed completely.

IN THE MOMENT

Think about the way you were taught math. Was the context in which it was taught (school culture, class environment, pedagogy, educational philosophy, etc.) conducive to enabling you to build a good relationship with math? Why? Bonus: You can engage your students in a similar discussion by asking them what factors might make them feel better about learning math!

And then there was my own complicated relationship with myself. I'll save the bulk of it for my eventual sex-drugs-rock-n-roll memoir, BUT let's just say that I've had a wild ride trying to figure out who I am through layers of imposter syndrome, perfectionism, OCD, depression, anxiety, and addiction. I feel so unbelievably lucky because this journey of self-discovery has forced me to stretch far beyond my comfort zone in an effort to heal, and I have tried SO many things—from self-improvement techniques to spiritual approaches to mindfulness and so much more! As I have expanded my toolbox with ways to help myself heal, those same tools have seamlessly made their way into my teaching practice as a way to help *students* heal. I truly believe that, at its core, education IS a form of healing. I have always brought my whole self to my teaching process. I treat my students like the messy, fascinating, complicated human beings that they are. I don't ever expect them to be one-dimensional caricatures of what "students" are supposed to be, just as I don't ever dare show them a one-dimensional caricature of what a "teacher" is supposed to be. My practice has always been a hodgepodge of everything I've learned in my own healing journey.

Education has been missing out. Instead of embracing the vast bodies of knowledge we have at our disposal and including all that we know about mental health, brain science, psychology, and so much more, we have instead shoved that all aside in favor of a "curriculum-first" approach. The result is that we are starting to see the consequences: teacher burnout, a lack of diversity (both inside math classrooms and in fields related to math), and increasingly disengaged and anxious students, to name a few.

We Are (Probably) Not All Licensed Therapists

Let's start by talking about the elephant in the room: the T-word. I get why some of you might be skeptical about trying any strategies that involve the word "therapy"—it can feel like a LOT of pressure, and also, the concept of therapy comes with baggage. But don't worry and try not to prejudge the process: This is something you *can* do! And it will help your students!

For me, Math Therapy started somewhat informally—as did my foray into actually teaching math, come to think of it! My first "students" were my Grade 12 classmates. Shortly after I had my own mathematical awakening, I decided that I wanted to help my friends feel the same freedom that I felt in discovering that I was capable of more math than I was led to believe I possessed the capacity for. After school, my classmates and I would take over the sketchy Coffee Time donut shop down the street, spilling our school supplies all over the burgundy, vinyl-topped tables, excitedly puffing away over ashtrays overflowing with cigarette butts (don't @ me, it was a different time!). It was a total vibe. I would walk from table to table, helping my friends understand how to do this or that, encouraging them to work together, giving them pep talks when they were feeling stuck, helping them work through their frustrations by taking deep breaths. In fact, aside from the cigarette smoke and lack of vertical non-permanent surfaces, it was a sight that would have potentially made Peter Liljedahl (aka the Godfather of Thinking Classrooms) VERY proud!

What I was really doing was helping my friends *build a better relationship with math*. I wasn't just teaching them content; I was helping them work through their deep-seated beliefs that they couldn't do it, I was teaching them mindfulness techniques for breaking through their perceived barriers, and I was showing them that learning math didn't have to look or feel a certain way. They were finding themselves as math learners, and even though every one of them may not have snagged an A+ in Grade 12 Finite Mathematics, they all voluntarily clamored into that Coffee Time every day, they each felt valued for their contributions to the mathematical conversation, and we all graduated feeling way better about our math abilities—and our*selves*—than we had when we had walked in 4 months prior. Some of my friends even went on to pursue degrees in postsecondary programs related to math like commerce and accounting—fields they never felt they belonged in until then. There was magic in that Coffee Time, and that magic was the stuff that Math Therapy was made of. I didn't know it then, but what I was doing was actually in line with what research would soon show: Interventions targeting affect and mindset can have an undeniably positive impact on students' math identity and resulting math performance (Boaler, 2013; Harackiewicz et al., 2012).

Over the decades that followed, I added a little structure; subtracted a LOT of cigarette smoke (ALL of it); went from a dingy donut shop to building a thriving tutoring studio of my own (instead of vinyl-topped tables, I opted for pink velvet couches, and instead of ashtrays, I regularly burn silver trays of lightly scented incense); and formalized my teaching process so there was a distinct method to Math Therapy (see Figure 2.3).

(see Figure 2.3)

FIGURE 2.3   THE MATH GURU (IN MY TUTORING STUDIO)

SOURCE: Racheal McCaig

Math Therapy isn't therapy in the clinical sense of the word, of course—I am not a licensed therapist and (probably) neither are you! But therapy takes many forms (retail therapy, speech therapy, physiotherapy, art therapy, therapy dogs). Don't get me wrong: Therapy with a licensed professional can be life-changing, and I personally have benefited from working with several therapists over the years. In fact, for deep-trauma work, I would definitely recommend seeking a licensed professional, but I want to make it clear that Math Therapy isn't that kind of therapy, and therefore, it doesn't require certification as a therapist.

 IN THE MOMENT

Are there practices that you consider therapeutic in your own life? Maybe it's a long bath, nature walk, cooking, or binge-watching home makeover shows. What qualifies them as therapeutic to you? Thinking along these lines will help you get comfier with the idea that therapy doesn't have to be traditional and that you can cultivate your own unique way to create therapeutic experiences for your students in a way that feels authentic to you!

Who Is (or Can Be) a Math Therapist?

I have this friend, Kaitlyn, who INSISTS she doesn't want to be in a relationship. Fine, except the thing is that Kaitlyn is always IN a relationship.

She just doesn't call it that. Currently, she's been "hanging out" with a guy basically every day for well over a year. They go on dates, spend holidays together, do basically *everything* together, aren't "allowed" to make out with other people—I'll stop there, but my point is, while I know that relationships can take many different forms for many different people, I feel as if we can likely all agree that Kaitlyn and this dude are *in a relationship*, right? But the thing is, if *he* calls it that, *she* doesn't want to be in it. But because *she won't* call it that, *he* feels insecure all the time. Labels are powerful!

Why am I bringing this up right now? Well, because while I came up with the idea to formalize and label Math Therapy, when I talk to most educators, they'll come right out and admit that it feels like half of what they do in the classroom *feels* like therapy ALREADY! And I say that to make YOU feel better (in case you weren't feeling 100% great) about the fact that you're likely way more equipped and positioned to call yourself a math therapist than you might think!

At its core, Math Therapy is a process that recognizes that mindset and everything that goes along with it is half (if not more!) of the battle when it comes to building a better relationship with math (Boaler, 2016). Math Therapy is a process that was built with the understanding that *we don't know everything* and it's okay (even encouraged!) to ask a specialist! When you sprain an ankle, you go to a doctor. When your tooth hurts, you see a dentist. When you feel like you just can't get out of bed in the morning because you have a math test later that day, you talk to a _____? Answer: math therapist!

As I mentioned at the beginning of this book, I believe that, in one way or another, we are *all* teachers of math. At the very least, we are all teachers of *math attitudes*—the way we communicate how we *feel* about math is influential to those around us, whether we notice it or not; and for those of us who are actually *teachers of math content*, responsible for not only translating math content, but in doing so, sometimes inadvertently sharing our own attitudes around how we feel and what we value when it comes to math. If you're in the latter group, you've likely dealt with your share of math meltdowns in the classroom and you know as well as I do that whether teacher's college prepared you for it or not, your time in the classroom is spent doing far more than teaching students how to think mathematically. In fact, if you're like most teachers, the fraction (math pun intended) of time you spend teaching kids actual math is equal to or less than the fraction of time you spend helping kids sort through their complicated feelings *about* math. Whether you think of it this way or not, you're already engaged in some elements of Math Therapy!

HOT TIP

Recent research has shown that math anxiety CAN BE contagious! If you experience math anxiety, the best thing you can do is be honest with your students about how you're feeling. The key is to make it clear that math anxiety is normal, that it has nothing to do with ability, and that it's something that can slowly fade away as we learn more tools and strategies to manage it!

As educators, we are uniquely positioned to carry out Math Therapy because teaching math already puts kids in a vulnerable place, whether we care to admit it or not. If we choose to look past that, not only are we missing an opportunity to help students heal and transform, but we risk retraumatizing students by failing to acknowledge their lived experience. As teachers, we are likely among the only people in our students' lives who understand how their relationship with math has unintentionally contributed to their sense of self, as well as one of the only adults in their lives with the understanding of how to begin to rebuild that relationship. You're already there. You're in the classroom, on the ground floor—you're already doing the work. Math Therapy is just another set of tools with which you can make that work more manageable, methodical, and meaningful.

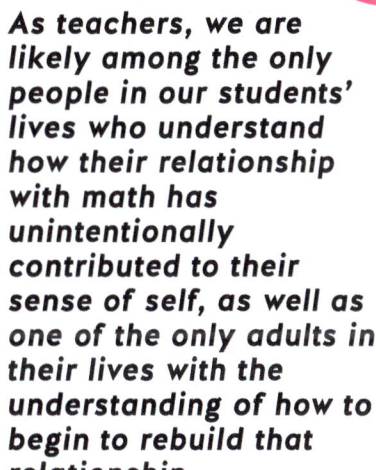

> *As teachers, we are likely among the only people in our students' lives who understand how their relationship with math has unintentionally contributed to their sense of self, as well as one of the only adults in their lives with the understanding of how to begin to rebuild that relationship.*

IN THE MOMENT

In what ways do you already feel like you're doing Math Therapy with your students? Make a list to remind yourself that you're likely more ahead of the game than you think!

WHAT PROBLEM ARE WE TRYING TO SOLVE?

I don't know about you, but I find that one of the things my students struggle with the most when solving word problems is actually deciphering what problem it is that they're trying to solve. It's like they get so excited about getting to the final answer that they forget that to find a solution we need to know what the problem is that needs solving in the first place! I'll be honest—that's how I tackle A LOT of my own life issues (I'm a Gemini, what can I say?!), which is why I do things like stop eating gluten in order to feel better, only to realize a year later that I still feel like sh*t and had I just taken a food sensitivities test in the first place, I would have realized that it's *actually* NUTS that I'm allergic to! Anyway, all that to say that before we start using Math Therapy to solve all our problems in the classroom, it's important to be clear about what problems Math Therapy can be used to solve!

Let *x* = Math Trauma

Remember, Math Therapy is a holistic process that brings together mindfulness-based tools and techniques that practitioners have been using in fields like counseling, self-help, and every possible genre of coaching for decades—and applies them to *math education*. I know it might seem like mindset is all we talk about these days, but the idea of actually incorporating any sort of thoughtwork into education is relatively new and a massively untapped resource.

Our collective goal when employing Math Therapy is *to empower students to build a better relationship with math, and in doing so, with themselves!*

I want you to pause and really let that sink in.

Note that the goal is *not* to propel every one of our students into a math-related career or to get them to be A students in math or to get their parents off our backs (JK!) or even to dazzle them all so mind-blowingly with the wonders and joys of mathematics that they become forever enamored with the subject. Our goal here is all about relationship-building. As I mentioned in the Preface, it is often through our relationship with math that we begin to develop a sense of worth. I recently interviewed math educator and mindfulness coach Deborah Peart for my podcast (Vakharia, 2023a), and I will never forget an anecdote she shared. She told me that, as part of a math intervention she conducted at an elementary school, she asked students to answer the prompt "Math makes me feel _____" on a piece of paper. Upon opening up all of the responses, she saw what one child had written:

Math makes me feel *like I am a good kid, when I do good at math.*

This student had articulated what many kids feel: that being "good" at math defines their worth as a student and as a *person*.

HOT TIP

Pay special attention the next time one of your students says they're not "good" at math. Ask them what they mean when they say that, and make sure it's clear that while math is for everyone, the way they feel about their math skills *now* is transient and will change, but more importantly: Our math ability is just one small, teeny part of our entire personalities—it's not everything, in the scheme of things!

The immediate goal of Math Therapy is to unpack math trauma as the starting point to healing one's relationship with math. As I mentioned when I shared the 5 M's of Math Therapy earlier in this chapter, it is unpacking that trauma that serves as a jumping-off point to helping us understand our math stories, and subsequently, it is those math stories that allow us to take a deeper look into what we believe about what we are worthy and capable of. Once we help our students face their math trauma, we open the gateway to be able to do several things:

- Empower students to learn to embrace the struggle necessary to solve math problems *in* the classroom—and to solve problems *outside* the classroom

- Equip students to take risks *in* math class—and *in* real life

- Encourage students to develop a creative approach to problem-solving *in* the classroom—and *outside* the classroom

- Energize students to believe they are capable of exceeding their expectations both *in* our classrooms—and *in* the pursuit of their wildest dreams

The goal of Math Therapy is *to empower students to build a better relationship with math, and ultimately with themselves!*

Speak it into being. Write it down somewhere you can see it every time you pick up this book to keep learning or set out to put Math Therapy into action in your own life. Keep this goal in your mind's eye every time you practice Math Therapy.

I know I am being really extra about this, but honestly, the importance of keeping goals in mind is kind of like that old saying about a fish climbing a tree or whatever. Do you know what I'm talking about? I just looked it up, and it's widely attributed to Einstein (but it probably wasn't him!).

> *Everybody is a genius. But if you judge a fish by its ability to climb a tree, it will live its whole life believing that it is stupid.*

> —*Probably not Albert Einstein*

The point is that if you judge the effectiveness of Math Therapy by whether or not your students all end up nailing a 4.0 in your class, you're going to be sorely disappointed. BUT if you judge its effectiveness by measuring the incremental improvement in your students' relationship with math, then you're likely to be pleasantly surprised, even perhaps low-key ecstatic. For real.

What Do Lizzo and Amelia Earhart Have in Common?

"Hey, Vanessa, you might not remember me, but 2 yrs ago, you visited my Grade 10 math class to talk about how to become friends with failure in math class! I just wanted to let you know that it really impacted me and that I had never seen myself as a math person until that day. I always thought that because I made a lot of mistakes that it meant that I could never actually be good at math, but your talk made me realize that mistakes are a part of being a good scientist and that learning from mistakes actually helps you get smarter. I never knew that before! Anyways, you motivated me not to give up and to follow my dream of becoming a doctor one day! Today I'm writing to tell you that I just got into a science program at Queen's University, and it's all because of that talk you gave my class, so thank you."

—*Jessica* (from my Instagram DMs)

As this DM from Jessica illustrates, doing mindset work with your students can have the effect not only of altering their view of themselves as learners in *your* math class, but of changing their entire sense of self, far beyond the walls of any classroom. In Jessica's case, I literally visited her classroom *once*, which is absolutely not ideally the way that Math Therapy—or mindset work of any sort—should be carried out to actually be effective. But even that 1 hour I spent in Jessica's classroom was the catalyst for her change in mathematical mindset. What made that hour even more effective was that her teacher then reinforced my message by engaging in one of my favorite Math Activities: the creation of a classroom failure wall! I'll tell you more about how to do this when we get to Chapter 3, but for now, all you need to know is that a failure wall is essentially a wall of images of folks who have failed on their way to success. So for example,

- Michael Jordan didn't make the varsity *high school basketball team on his first try* (granted, he was a first-year at the time!).
- Elliot Page was told that they would never be a successful actor.
- Steven King's first manuscript was rejected by 20 publishers.
- Walt Disney was fired from his job as a newspaper reporter at age 22 by an editor who said he "lacked imagination and had no good ideas."
- Oprah Winfrey was once fired by a news producer who told her she was unfit for television news.
- Bill Gates's first business venture was a total failure and went bankrupt.
- Salma Hayek was discouraged from pursuing a career in film because someone told her she wasn't right for the business.
- Katy Perry was dropped from *three* different record labels.

I could go on, but you get the point!

Jessica's teacher threw up a bunch of images of celebs with similar stories on the wall, so that every day her students would see their faces and be reminded of the message I had given them in that 1-hour talk!

In some cases, Math Therapy will happen as a series of math-adjacent, noncurricular tasks (for example, as in the case of my visit to Jessica's classroom, you might spend several lessons teaching your students about failure and imploring them to explore how failure has played a positive role in their lives). In other cases, Math Therapy will make its way into your curricular tasks (for example, you might have students show off a failure or mistake they made while engaging in a particular math task, exploring what mathematical lessons they learned in the process)! Finally, you will weave the steps of Math Therapy not only into your practice but also into your physical space so that you cultivate a culture in which students are consistently building a better relationship with math—otherwise known as a Math Therapy classroom!

WAIT, HOW LONG WILL THIS TAKE?

Let's just get to the thought I know you're all thinking right now: None of us has time. For anything.

I hear you, I promise. But if you picked up this book, I'm guessing it's because the title spoke to your soul, which is likely screaming something along the lines of, "AHHHH MY STUDENTS ARE STRESSED AND CRYING AND STRUGGLING AND I DON'T KNOW WHAT TO DO."

Am I right-ish?

Remember in the last chapter when we talked about the idea that when we're in an anxious state, we actually *can't* learn anything? All of our mental resources are so tied up in trying to help us manage the anxiety we're feeling that someone could literally say, "I spoke to a legit psychic, and I know for a fact everything will be fine," and *we wouldn't hear them*. Or we would, but we wouldn't be able to digest that info. It wouldn't matter. THAT is how our students are feeling. So even though we feel like we don't have the time, we are actually already spending SO much of our time managing anxiety-related behavior like lateness and chronic absenteeism, putting out fires, wiping tears, and trying to console our struggling students. I am so, so sorry that you're dealing with all of that. It's hard, and it sucks. And it sucks even more that I'm telling you to do yet ANOTHER thing, but I promise, if you do this thing, all of that OTHER stuff will take up less of your time and resources because it will start to happen less and less. There's a fitness metaphor in here somewhere: This is like that person who swears they never have the time to work out, so they don't, and they start developing chronic pain from lack of movement and getting sick all the time from lack of cardio, so they spend all this time going to the physio and doctor and not being able to do anything because they feel like sh*t, only to realize that if they had just invested the time up front to EXERCISE IN THE FIRST PLACE they would be saving a ton of time in the long run. And by "this is like *that* person," what I really meant is "this is like . . . *me*." Guilty as charged!

The other thing I want to say is this: Changing our beliefs takes time. I know that seems rather obvious, but I want you to just think about this for a second. We are all a product of the habits and beliefs we've built over our lifetime, and the two aren't mutually exclusive but rather intimately related. We may not think about it this way, but a student who walks into math class without their homework completed day after day has a habit of not completing their homework, day after day. Now, that may result from the belief that there is no point in DOING their homework because they suck at math, and of course, they will likely perpetually continue to *feel* like they suck at math as a result of never doing their homework. It's a vicious loop, and we're all familiar with it in our own way, in our own lives. But you can help your students with Math Therapy to change their habits AND their beliefs about their math ability—that's the whole point—and the 5 steps are full of strategies to help students slowly rebuild their relationships with math. I say slowly because most of your students are showing up with a lifetime of belief and habit baggage, and it's going to take a while to change them, straight up. Just think about any time you have had to build a new habit: working out, flossing your teeth, meditating every morning . . . at first, it's a struggle and it doesn't feel natural. Maybe you even broke your streak and needed to restart from the beginning. Maybe you gave up, only to come back to it months later.

The relationship between beliefs, identity, and habits is rooted in behavioral psychology, and in his book *Atomic Habits*, Clear (2018) emphasizes the importance of identity-based habits, where you focus on changing your beliefs about yourself to align with the habits you want to develop. Simply put: The key to building and sustaining a habit is to believe that you are the type of person to whom that habit belongs. For example, if you don't give a crap about dental care, your tooth-flossing habit isn't going to stick. But if you become someone who believes that dental care is important and that you can personally affect the longevity of your gums, you're more likely to keep up that daily flossing even when it's inconvenient or gross or annoying. Similarly, if your students *believe* they really can build better relationships with math, they're more likely to take on the habits (e.g., showing up to class with their homework at least *attempted*) that can get them there. That's why it's important to be patient, to start with mindset work (hello, Step 1), and to understand that undoing a lifetime of habits and beliefs doesn't happen overnight. Which leads me to . . .

Beyoncé Wasn't Built in a Day

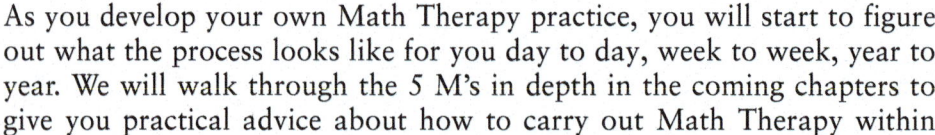

As you develop your own Math Therapy practice, you will start to figure out what the process looks like for you day to day, week to week, year to year. We will walk through the 5 M's in depth in the coming chapters to give you practical advice about how to carry out Math Therapy within your classroom. But for now, I want you to know that Math Therapy isn't just something that you do once and get it over with. It's something that will be woven into the fabric of your teaching practice, every moment of every day!

> *I want you to know that Math Therapy isn't just something that you do once and get it over with. It's something that will be woven into the fabric of your teaching practice, every moment of every day!*

How Long Should It Take Me to Go Through the 5 Steps?

It is always best to go through one round of the process sequentially and to do so as quickly as possible. I have found that 2 weeks is enough time to get through the 5 steps in most scenarios—but you will determine what works for you in your particular classroom (or in your small coaching group) in your particular set of circumstances. The important thing is to work sequentially and give students time to process and discuss. It is within these discussions that the greatest growth can occur. Give your students time and space—and keep at it.

Not to be discouraging, but in my experience, Math Therapy is never really *complete*. Like, imagine you asked your therapist when you would be done being therapized??? They would be, like, "Ummm, never?" We ALWAYS have something to learn about ourselves until literally the day we die. It's the same idea when you're on a mission to heal your math trauma and build a better relationship with math! You can make consistent progress, but if you're looking to be "done," you're never going to feel successful. I mean, I have been on A JOURNEY with math, and while math and I *are* bffs, I still have moments when math anxiety takes over (e.g., tax season) or when I feel like I'm not good at math (e.g., when I'm at a math conference and I'm like, Am I the only person in this room who doesn't quite know wtf is going on?). The question isn't whether in the first round of Math Therapy a step is "complete," but whether you sense that you have provoked your students to think about their relationship with math in a *different* way. Each step takes your students further along the journey of *beginning* to heal their relationship with math, and if you can move your students from step to step, they *will* start to feel different.

Once you have tapped into the Activities for each step, you can move on to the next step. Just remember that one-and-done is not the goal.

The goal is to give your students enough fuel to continue to ignite their relationship with math that they're curious and open to investing in the process. And as I've said before—but I'll say again because it's THAT important—in a Math Therapy classroom, Math Therapy happens every day!

Do I Have to Start With Step 1?

Yes.

Let me ask you this: Imagine I told you that I knew 100% for sure that next year you were going to win $10 billion. Like, I knew that WITHOUT a shadow of a doubt. And then I was, like, "But still, you should watch your spending, save for retirement, and make sure not to take vacations or do anything extravagant until then. And for sure don't quit your job even if you want to because you need to make money." Wouldn't you be like . . . but why should I bother? You just said I'm going to be RICH AF, why would I do all of these things that make NO sense to do if I'm going to be LOADED? That's exactly how your students feel, probably right now. What they're thinking is, *I can't get better at math because I'm not a math person. So why bother trying or even thinking about whether I've had math trauma or not; none of it matters because I'm unable to get better at math. So . . . no thanks.*

THAT is why Step 1 MUST COME FIRST. We MUST mythbust students' misconceptions about their abilities; otherwise, none of the other steps will land. No one wants to get out of bed at 6 a.m., save every penny, and stress about their 401(k) when they know $10 billion is about to land in their bank account, you feel me? If you're still skeptical that you absolutely need to start with Step 1, please read what I said, like, 2 seconds ago, in the "Wait, How Long Will This Take?" section. Convinced?

 **IN THE MOMENT**

How much mindset work have you already done with your students? Don't worry if the answer is "none"! Take an inventory of what you think your students might already believe when it comes to abilities being fixed or fluid. This will be a good starting point for you when you get to Step 1!

A Note on Classroom Culture

Last week I got an email from a teacher that went somewhere along the lines of this:

> *Vanessa, I loved your Math Therapy course, and I am so excited to try out some of these strategies in my classroom. The problem is, I don't know how to even start a conversation about Mythbusting with my kids because whenever I try to get them to discuss anything, it turns into a war zone! I'm serious—they start throwing things and ripping posters off the wall and making nasty remarks to one another any time I try to get them to collaborate or engage in group discussion. What should I do?*

This teacher legit used the words "war zone" and referenced "tearing things off walls." This is real. And I bet some of you are thinking, "Yep, been there, done that!"

HOT TIP

Bring humanity into the classroom! If you haven't already, start priming your classroom with empathy and compassion. Create a classroom agreement that makes it clear student voice is valued! Start your class with a mindfulness practice each day! Teach your students conflict-resolution skills! Teach your students about the six basic types of emotions (happiness, sadness, fear, anger, disgust, and surprise) and explore how to identify them! Most importantly, get to know your students and make sure they know their voice is valued in your classroom.

So in that situation, how can you move forward with Math Therapy? My friend and education expert, Deborah Peart, once told me,

> *Mindfulness practices don't fix a toxic culture. It fits within a culture grounded in humanity.*

Math Therapy isn't a magical cure for everything. It is a way to help students heal their trauma and build a better relationship with math. In order to even begin to do that, a *culture grounded in humanity* needs to exist. There are many tips in this book for how to create a Math Therapy classroom by working on classroom culture, but I caution you to make sure that humanity, empathy, and compassion are already present—and if they aren't, to work on that first. It will pay off in SO many ways, I promise. Some ways to get started on fostering these three essentials are by modeling kindness yourself, teaching empathy and perspective-taking, helping students identify and deal with strong emotions, and partake in empathy-building activities, like storytelling and role-playing, that help students understand diverse perspectives.

Check In With Yourself!

How are we all feeling? Excited? Scared? A little bit of Column A and a little bit of Column B? Whatever you're feeling, I promise that it's totally normal. We all know that new things can feel scary—and sometimes even impossible—at first. Just imagine when Henry Ford was like, "Guys, don't worry, I've got this thing on four wheels, and we can all just collectively move around at super high speeds, and everything will be JUST FINE, I swear!" It always feels safer to stick with what we know than to try something new, but if you're an educator reading this, you know as well as I do that the "old way of doing things" has been failing a lot of our students and that we desperately *need* a new approach or else we run the risk of retraumatizing students for the rest of eternity. If you have this book in your hot little hands right now, I'm pretty sure that's the last thing you want.

BEFORE WE MOVE ON

I know you're excited to move on to the next chapter—I'm excited for you to do it too! But first, I have a couple of treats and questions to help you reflect, process, and CELEBRATE your Math Therapy journey!

Treat Yourself

You just finished a WHOLE other chapter, and it's time to treat yourself to a little therapizing in whatever way suits you best! One of my favorite ways to engage in classic talk therapy is by talking . . . to someone I REALLY feel like talking to. Take 10 minutes to catch up with an old friend, call someone you haven't spoken to in years, or hop on a Zoom call with someone you miss. There's nothing like a good conversation to spark new ideas when you're in the middle of processing new information (like this entire chapter, for example)! Take a few minutes now to catch up on the latest gossip with whomever you feel like talking to, and I'll see you in the next chapter!

ASK YOURSELF

Get out your notebook, notes app, voice memo device, or your Math Therapy journal (remember you can find pages to download on the Math Therapy site) and reflect on the following:

1. Given your lived experience, what personal touches do you think you might bring to Math Therapy to make it your own unique practice?

2. Can you think of one of your students (or multiple students!) who might benefit from Math Therapy? Why?

3. What does "building a better relationship with math" mean to you?

4. Which step of Math Therapy excites you the most? Which one scares you the most? Why?

5. Do you feel like your current classroom culture is conducive to Math Therapy? If not, what needs to be done?

6. Consider your classroom culture and what it would take for it to become a Math Therapy classroom. Which step do you think will be the most impactful for your students?

⊕♡π

CHAPTER 3

MATH THERAPY STEP 1

Mythbust Mindsets!

THE 5 M'S OF
MATH THERAPY

✓ STEP 1: Mythbust

STEP 2: Moderate

STEP 3: Motivate

STEP 4: Makeover

STEP 5: Measure

In this chapter, we get to:

1. Discover additional important information about growth mindset

2. Understand why what our students focus on plays a huge role in shaping their relationship with math

3. Dig into Step 1: *Mythbusting*

4. Stock up on concrete Actions and Activities to bring Step 1 of Math Therapy to life in our classrooms in an immediate and impactful way

FIGURE 3.1 CHRISTOPHER HAVENS

On Season 3 of the Math Therapy *podcast, I interviewed a man (Christopher Havens) who was in prison for murder (Vakharia, 2021). I interviewed him straight from jail. He had to use a pay phone to call us, and because he only had a 15-minute talk-time allotment, he had fellow inmates line up behind him so that when his 15 minutes were up, they would dial us and pass the phone to Christopher so that we could continue the interview. We had to do this seven times to complete the full interview!*

Christopher (see Figure 3.1) was in the midst of a 25-year murder sentence when, out of nowhere, someone slipped a packet of math puzzles under his door when he was in solitary confinement. Though he never passed high school, Christopher found himself captivated by the puzzles. One thing led to another, and today, not only has Christopher had his research published in an international journal for advanced mathematics, but he has also founded the Prison Mathematics Project, an organization dedicated to helping prisoners rehabilitate through math. And he does all this—from jail—where he is serving out the rest of his sentence.

When I asked Christopher what it was that made him fall in love with mathematics, he told me that it was the meditative feeling he fell into when his brain was working on solving a problem and the subsequent feeling after the fact—that the act of solving the problem had somehow changed him. For someone who's a part of a population that North American society has essentially deemed unable to change or rehabilitate in any way, I can't imagine a more powerful, moving, and literally life-altering experience.

When I asked Christopher if he had heard of growth mindset, he said no. When I asked him if he thought that anyone could learn mathematics, without hesitation, he yelled, "Absof#*ckinglutely."

To hear my full conversation with Christopher, listen to *Math Therapy*, S3E10, "How Math Rehabilitated a Murderer" (https://bit.ly/3URfYQa). Also, I have to tell you guys that last year, the *Math Therapy* podcast was officially added to the Securus system, meaning that inmates across the United States can listen to the podcast for free, and I get emails monthly from prisoners telling me that they have totally changed their view on math. It's probably one of the coolest things to ever happen to *Math Therapy* and just goes to show that Math Therapy can literally happen ANYWHERE!

Notice how Christopher came to an understanding of the power of growth mindset through his own personal experience and keep this in mind throughout the chapter.

STEP 1: *MYTHBUST*!

You have made it to Step 1—let the games begin!

MYTHBUST

The first step of Math Therapy is all about how to mythbust your students' misconceptions about what it means to be "good" at math, what it means to even do math, and what it means to learn and grow in meaningful ways, both inside and outside of the classroom. In this step, you will make use of loads of classroom strategies to convince everyone in the room that we are all capable of doing literally anything we put our minds to—STARTING WITH MATH!

Why It Works: Mythbusting is the first official step of Math Therapy because if we don't put dedicated effort into helping our students cultivate growth mindsets, it is unlikely that any of our amazing strategies or incredible content will ever have the chance to reach them.

Over the years, I have had so many educators tell me that, even though they're super positive with their struggling students and reiterate the message that there is no such thing as a "math person," their students don't believe them. Saying something over and over isn't always the best way to get a message across. The best way is usually to let someone have a set of experiences that allows them to come to their own conclusions about whatever it is you're trying to convince them of. So if you're trying to tell your students that their math abilities can grow and change because their literal brains can grow and change, you have to give them opportunities to discover that ALL on their own. THAT is what mythbusting is all about!

How It Works: The Step 1 *Mythbust* Toolkit will help you equip your students with strategies to explore growth mindset on their own terms through math activities, noncurricular tasks, and reflection strategies. It will also equip *you* with techniques and strategies to ensure that these activities take place against a growth mindset backdrop. That means cultivating a classroom environment where mistakes are normalized, effort is celebrated, and students feel empowered to take on challenges and take ownership over success.

> *Cultivate a classroom environment where mistakes are normalized, effort is celebrated, and students feel empowered to take on challenges and take ownership over success.*

There are three mega misconceptions we're going to bust in this step:

1. The myth of learning: This is where we mythbust the idea that our abilities are fixed!

2. The myth of failure and success: Here, we mythbust the idea that mistakes and failure are bad things!

3. The myth of "good at math": Finally, we bust the idea that there is only ONE way to be good at math!

Like all 5 steps of Math Therapy, *Mythbusting* isn't a one-and-done technique but instead is a set of ideas and strategies that will both help you set the tone of your classroom while also enabling you to *build* and *maintain* a Math Therapy classroom. Some of the tools in your toolkit will be things that you can do *once* with your students, and others will need to be used *consistently* and *repeatedly* with your students.

Now, before we get to building your *Mythbust* Toolkit, I want to share two of the most important growth mindset concepts that have completely changed my teaching practice! The first has to do with how our thoughts shape our reality, and the second has to do with how our brains can grow and change as a result!

IN THE MOMENT

Stop and ask yourself what you think it means to be "good at math." Now, ask your students. Do your definitions line up? If not, engage in a discussion about some of the myths around what it means to "do math" and "be good at math" to help your students see that there are SO many ways to *do* and *be good at* math!

GROWTH MINDSET 2.0

I know, I know—you're probably sick of hearing about mindset (Clear, 2018; Dweck, 2007; Kahneman, 2011), but I promise I'm going to tell you something you don't know. Why? Because there is a very specific way in which I want us to approach growth mindset when thinking about Step 1 of Math Therapy. I want us to think about it in a way that will empower our students to take growth mindset from a concept—to an irl *way of being*.

Angel Numbers

Okay, I'm going to be honest: I'm one of those people who's into angel numbers. Now, you're probably either nodding your head in agreement, as in, "Me, too, babe," OR you're scrunching your face up, as in, "wtf is an angel number?!" For those of you in the latter group, let me just explain that an angel number is a specific sequence of numbers that is believed to hold spiritual or mystical significance or to just be plain lucky. Usually, these numbers consist of repeating digits like "333" or "1111" (that's why everyone's all, like, "OMG, 11:11, make a wish!"). Some people believe that certain numbers mean certain things, and the internet is full of suggestions about what specific number sequences mean. I'm literally wearing a gold necklace with 222 engraved onto it and to me, that symbolizes intuitive knowing and guidance—I'm in deep, what can I say!

Now, I'm bringing this up at the risk of you thinking I'm a total whack job because I'm about to make a very important point about how our brains work. You see, the whole theory around angel numbers is that, when you *see* them, they're a sign that you're in the right place, you're being guided, and that everything is going to work out. So when I look at the clock, for example, and see 2:22, I'm like, OMG, IT'S A SIGN! Same thing when I pass a house with 222 in the address, my restaurant bill comes to $22.22, or even when I see that my inbox has reached 222 unanswered emails (actually, that sends me into a state of panic, but that panic is mitigated by the feeling that MAYBE IT'S A SIGN)! There is NO limit to how excited I will get over seeing repeated 2s; in fact, I'm so crazy about the whole thing that I may or may not get a 222 tattoo—I just haven't decided WHERE yet. (tmi?) [*P.S.* It is now 3 months after I wrote that sentence, and I now have 222 tattooed onto my upper left arm, soooo . . .]

The point is, not only do I get excited every time I see those repeated digits, but I am convinced that I see them more than the average person. And guess what? I probably *do*. Now, as much as I love to believe in signs and synchronicities, I know that the reason I likely see repeated 2s say, more than you do, isn't necessarily because of some angel sending me a message . . . it's because of science. Specifically, it's because of brain science—and more specifically, it's because of my reticular activating system (RAS; Rothstein & Stromme, n.d.).

HOT TIP

Did you know that there are a few things that will usually *always* get the attention of our RAS? Those things include personal relevance (information that is relevant to personal goals, interests, or immediate needs), contrast (information that stands out), and surprise (new or unexpected information). Incorporate those elements into the steps of Math Therapy to really grab the attention of your students!

Our RAS acts kind of like the security detail of our brain. Just like Britney Spears's security detail always made sure that only the *worthiest* guests made it through to Brit-Brit while the rest were told to F-off, your RAS sorts through billions of bits of information per second and organizes it into piles of stuff that are worthy of your attention and mounds of trash that can be immediately discarded. As teachers, we're always on the lookout for ways to make sure that our lessons and words of wisdom (about growth mindset, for example) actually make it through our students' super-picky RAS filters. But how do we do that?

Well, if I had a simple answer for you then this book would for sure sell a trillion copies and I'd win a Nobel prize because, as educators, it often feels like the bane of our existence is trying to figure out how to get our students to pay attention to us! So no, the answer isn't simple, but I CAN explain a very important piece of information about how our RAS works. Let's go back to Britney Spears for a sec. How did her security detail know who to let past the velvet ropes or not? Well, for the most part, Britney (or her controlling parents) would tell the security folks who to look out for—maybe even give them a handwritten guest list. That way, they were laser-focused on spotting the important people that needed to be let through. In the same way, my RAS knows that I'm very interested in angel numbers, especially those containing repeating digits of 2. The second a set of repeating 2s enters my periphery, my RAS filter is like DING DING DING DINGGGGGGG and basically shines a spotlight on those 2s so that they stand out to me. And THAT'S why I'm convinced I see 222 more than say, you do—but the truth is that I probably just NOTICE them more than you do because of my RAS!

Our RAS is like that pesky algorithm on your smartphone—you know what I'm talking about! One second you're talking to your friend about how tired you are, and the next, your phone is showing you ads for energy drinks on Facebook, suggesting you follow crypto-bros-turned-energy-coaches on Instagram and spamming you with emails about how, for just $29.99, you, too, can get your energy back! Your RAS is so intent on serving you the content you want to see that if, for example, you're fixated on the idea that you suck at math, guess what it's going to show you? Proof that you suck at math!

IN THE MOMENT

Algorithm check! Hear me out: We talk *a lot* about algorithms in math class, but what if we used this as an opportunity to highlight what an algorithm does both *irl* and *in our brains*! This is an opportunity to teach your students about the RAS while also talking about when algorithms may or *may not* be helpful in math class. Some questions you might ponder include these: What is an algorithm? How does an algorithm behave? Is our RAS kind of like an algorithm? Why or when is that helpful? Why or when is that not helpful? Can you think of an algorithm used in this class? Why or when is it helpful? Why or when is it not helpful?

Watch Your Step!

I don't want to make assumptions, but I'm going to go out on a limb and guess that you've likely heard of *neuroplasticity* by now. When I bring up growth mindset to most of my friends, they're usually like, "Oh yeah, right, your brain is plastic and can change or something?" Next to understanding how our RAS filter works, neuroplasticity is probably *the* most important thing for you—and your students—to understand because neuroplasticity is the brain's ability to learn and change. As brain-based learning expert Liesl McConchie put it in our Math Therapy interview, "Neuroplasticity is a message of hope" (Vakharia, 2023b). It is the science behind the fact that the brain can create new connections between neurons and strengthen existing ones, helping us to adapt to new things and learn from experiences throughout our lives.

Now, I'm going to use a very Canadian analogy here. In the winter, sometimes there's so much snow that before the snowplow people can even get to removing it, you still need to walk down the street to, like, get to the bus or walk your dog or whatever. When you're the first person awake the morning after a huge snowfall, it truly is magical. But it's also very annoying because now you need to walk your dog, there's a blanket of snow that's 7 inches deep, and every step you take is a struggle because you're sticking your foot into 7 inches of snow, pulling it out, and then doing it all over again—imagine how hard that is! Now, if you're the 10th person to walk your dog that morning, you're in luck. By then, nine other people have put their feet into those same snowy footprints, so by the time you get out there, there's a trail of hard-packed snow for you to walk on, and it's really no struggle at all! (Also, if you're really struggling with this metaphor because you live in California or some other warm place, sub in "sand" for "snow.")

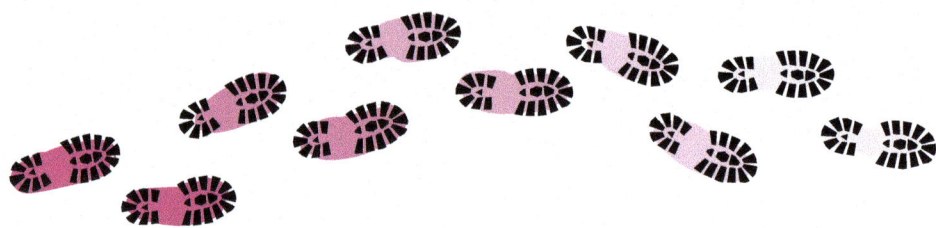

Our brains are basically the snowy streets of Canada in December! Or at least, you can think of them that way. Your brains CAN create a path (aka a new connection!) between your metaphorical house and the equally metaphorical bus stop on a snowy morning, but it's usually challenging at first. However, with repetition and consistency, that pathway becomes easier to travel, and hey, you might get so used to walking that same path that you're eventually able to do it without much effort or thought! It is this element of growth mindset that can be supercharged to help our students tap into their abilities to change the way they think about math. This is a way to think about both neuroplasticity as well as the consistency and repetition needed to solidify new connections, patterns, habits, and ways of thinking.

Find multiple ways to explain neuroplasticity to your students and to reinforce the message that *our brains are constantly growing and changing.* Here are some ideas:

- Read a book together. (*Bubblebum Brain* by Julia Cook and *Outliers* by Malcolm Gladwell are two of my faves!)

- If you love science, teach them about the brain science behind neuroplasticity.

- Share with them my footsteps-in-the-snow analogy (see p. 64).

- Use interactive brain models to allow kids to explore the brain and how it grows and changes as learning happens! (https://bit.ly/48zHl4j)

I think of neuroplasticity and the RAS as frenemies because they feed off each other for better or worse. Say your student's RAS is focusing on information that confirms that they're bad at math, which means their RAS is going to let in bits of information that reinforces this message. For example, their RAS is going to pay extra-close attention to a bad test score and only a little bit of attention to a better test score. Their RAS is going to go DING DING DING when they fail to get a math question right or when a parent acts disappointed in their report card mark or when they struggle to understand a new concept—and it's going to basically IGNORE info relating to the times they get a math question right, when a parent acts thrilled about their report card mark, or when they finally grasp that new concept! Now, the RAS will feed that info to your student's brain, and the brain will get to work on both building new connections *and* reinforcing them. That means it will continue to strengthen the belief that they suck at math, paving those footprints in the snow until your student just thinks that thought over and over again, subconsciously, without much effort at all. And THAT is why, as the old adage goes, What you focus on grows!

A caveat: *growth mindset is absolutely not in any way* an antidote to the systemic barriers that folks struggle with every day. This is not a "pull your-

selves up by your bootstraps" situation, and mindset is rarely the *only* thing standing in the way of our students and a positive relationship with math. We all understand that our students do not exist on an equal playing field inside— or outside—our classrooms. In STEM and throughout education, we are coming to realize,

> **We all understand that our students do not exist on an equal playing field inside—or outside— our classrooms.**

more and more, that students of historically and contemporaneously marginalized groups have not been given equitable access to education and that, for some, growth mindset may not come easily.

IN THE MOMENT

Go through your class list and consider the unique barriers your students might be facing. You might realize that you don't know much about some of your students and a lot about others. This is a good opportunity to fill in the missing gaps! Understanding what our students might be dealing with inside and outside of the classroom allows us to tailor our approach and show empathy as we take them through the 5 steps of Math Therapy!

It is important to ensure that while we're emphasizing the brain's universal neuroplasticity we are not claiming that all students universally have the same access to agency when it comes to activating, accessing, and internalizing that neuroplasticity. Educating ourselves about factors that affect *all* of our students will help us find unique ways to meet each of our students where they are, helping them each begin to cultivate mindsets that lean more toward notions of growth rather than fixed. While we as educators have little, if any, power in the systems that oppress many (including us!), we still can make a very real difference in the lives of our students through our words and actions every single day.

> *Educating ourselves about factors that affect all of our students will help us find unique ways to meet each of our students where they are, helping them each begin to cultivate mindsets that lean more toward notions of growth rather than fixed.*

HOT TIP

Watch your language! Our students' RAS may be on the lookout for language that reaffirms how bad they are at math. Watch the way you communicate both verbally *and* nonverbally to ensure that students are getting the message that they're better at math than they may think and that you really do value the pace at which they are growing and learning, regardless of what that pace may be. Look out for judgment terms like *good, bad, genius,* and *talented,* and replace them with growth mindset terms such as *progress, not quite there yet, effort,* and *improvement*!

PUT *MYTHBUSTING* INTO PRACTICE

At the beginning of the chapter, I shared Christopher's experience of myth-busting. In his case, he had always been taught that his abilities were fixed. On top of that, he had been incarcerated for 12 years when I spoke to him. I won't profess to be an expert on the American prison system, but I do know that many institutions operate under the belief that prisoners cannot be rehabilitated, which is simply code for *People can't change.* Trying to mythbust deeply ingrained ideas when surrounded by external factors that don't support our growth is not dissimilar to what many of us face in the classroom. In Christopher's case, that mysteriously delivered packet of math puzzles gave him the opportunity to challenge those deeply held beliefs, but he had to come to the realization that he could do the puzzles all on his own, without the guidance that our students are lucky enough to get from us. By engaging with the math, he began to feel that spark that accompanies learning. He began to be propelled forward by those "aha" moments. He began to *feel himself change.* It is important to note that he wasn't slipped a packet of math facts or algorithms or "I do, we do" exercises. He was slipped an envelope of puzzles that sent him a clear message: I believe that you are smart enough to figure these out. As I will discuss later in this chapter, providing ALL students, regardless of ability, with opportunities to feel challenged is actually *conducive* to their Math Therapy journey. It mythbusts the idea that our abilities are stagnant, that mistakes aren't "allowed," and that math is only for certain people.

For our students to really embrace and embody growth mindset, it takes more than a poster on the wall or a set of repeated affirmations (both of which I LOVE, by the way)! It's not enough to *believe* that growth mindset is a real thing; students need *see it for themselves*. In fact, it can actually be legit damaging to TELL a student repeatedly that they CAN CHANGE just by believing they can without giving them the opportunity to CHANGE THEIR BEHAVIOR to actualize that belief. This is the growth mindset mantra I live by whenever I'm deciding whether or not to use an activity or strategy with my students:

Believe. Behave. Become.

Believe you can be what you choose, **behave** like whatever you've chosen, **become** who you choose to be.

Our current beliefs about our own abilities are often a direct result of our past behaviors. For example, for a long time I believed that I couldn't quit drinking. I believed that because . . . well, let's just say I had a *complicated* relationship with alcohol, and it was what I turned to when I felt sad, happy, or in between. When I stopped drinking for a few days or weeks or even months, I would ultimately go back to it. So I believed that I could never stop; I believed that if I stopped my life would suck; I believed that it was just *too hard* to stop—that is, until I stopped drinking on October 7, 2018.

Now, I'm not going to get into every detail because I'm saving that for my eventual sex-drugs-rock-n-roll memoir, but this time, like all of the others, I tried yet another set of strategies. I knew what hadn't worked in the past, so I tried something else. I wasn't positive that this time would be successful, but I kept trying, adjusting along the way, using the lessons I had learned from my past failed attempts at getting sober . . . and guess what? Over 5 years later, I am sharing this story with YOU and can confidently say that I AM OVER 5 YEARS SOBER, BABY!!! Of course, everybody's journey is different, and what worked for me may not work for others. I can only speak to my personal experience here in saying that believing that I could change played a pivotal role in my recovery. It was that belief that empowered me to finally seek out the resources and help I needed to change my behavior, to become the person that I am today.

While obviously our stories are very different, Christopher went on—and is still on—his *own* "believe, behave, become" journey, and last I heard, there was talk about making a movie about him . . . and I swear the words "Bradley Cooper" were mentioned!

ME	
BEFORE	
Believe	I believed I couldn't stop drinking.
Behave	When the going got tough, I turned to alcohol.
Become	I became unhealthy and unhappy.
AFTER	
Believe	I believed that I could stop drinking.
Behave	When the going got tough, I tried new strategies.
Become	I became healthy and happy and realized I could change and grow into the person I wanted to be.

CHRISTOPHER HAVENS	
BEFORE	
Believe	He believed that his brain couldn't change.
Behave	When the going got tough, he turned to old habits and learned behaviors.
Become	He became hopeless and unproductive.
AFTER	
Believe	He believed that his brain could change.
Behave	When the going got tough, he was open to learning new strategies and behaviors.
Become	He became hopeful and productive.

Use this irl! Keep this before-and-after chart in your back pocket for the next time one of your students says they wish they _____ ("were better at math," "didn't feel so stressed," "could stop procrastinating," "could reach their goal" . . . and so on). Remind them that we can't just think our way into different outcomes; we have to change our behavior to *arrive at* those outcomes. Challenge them to fill out a similar table so they can see what is standing in the way of becoming the person they want to be and how they can change their behavior to get there!

Before we move on to our VERY FIRST TOOLKIT, I thought I would give you a little extra motivation to get into *Mythbusting* by sharing a story that an Ontario high school math educator, Jamie Mitchell, sent me! It's adorable, so buckle up.

"Mr. Mitchell, I think I found the perfect person for our conference keynote."

My class was working on a big project to design a STEM-themed conference for Grade 9 students in our school board. Mackenzie, a Grade 11 student in the I-STEM program, was pitching me on a speaker she had found.

"Her name is Vanessa Vakharia, and I saw her speak at Waterloo University last year."

Full disclosure, at the time I had no idea who Vanessa was, but I trusted Mackenzie, so I heard her out. Mackenzie raved about Vanessa's message: specifically embracing failure in math class while being your authentic self.

This was important for me to hear too. At this point, I had already been teaching Mackenzie for 3 years, so I knew that her experiences with math had been rocky before high school. Like all of us, Mackenzie is a math person, but like most of us, she had been told that she didn't fit the mold of a "typical math person." Mackenzie shared what I could only describe as bullying, mixed in with some outdated ideas around gender and who is allowed to be good at math. Mackenzie had a lot of math trauma upon entering Grade 9, and even though I didn't know it at the time, through the community we built in our math classes, I played a small part in helping her overcome that trauma.

Mackenzie graduated from high school in June 2023 (see Figure 3.2). My pride for her accomplishments and growth was only surpassed (I'm sure) by her own pride in meeting her goals, specifically in being accepted into Queen's University Engineering. Mackenzie and her peers were always going to be exceptional. My work was to make sure that they saw that exceptionality in themselves.

FIGURE 3.2 JAMIE AND MACKENZIE AT MACKENZIE'S HIGH SCHOOL GRADUATION.

SOURCE: Jamie Mitchell & Mackenzie Richardson

I realize now that the conditions I created in my math classroom helped Mackenzie and many more students like her. Every day, we embraced failure; we were not motivated by grades but rather by our understanding; and my steadfast belief in the ability of all of the students I taught created the fertile ground for where we can all thrive.

I worked hard to ensure that math class was always a positive space where every step forward was celebrated, and our mistakes were celebrated even HARDER. Students knew that I believed in them, even on their worst days. Math class became a safe space to show our humanity rather than a place of uncertainty to be avoided.

YOUR *MYTHBUST* TOOLKIT

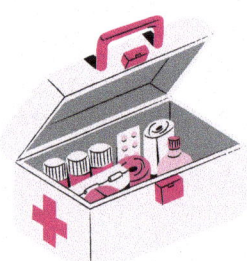

Why It Works: We have seen that an emphasis on shifting student mindset *does* correlate with a shift in math efficacy and performance, and we have historically seen that a superficial adoption of growth mindset *will not work* (Boaler et al., 2021). It is not enough to tell students to simply believe that their minds can change. This toolkit works because it isn't just all about lip service. Instead, it makes use of what we already know about the RAS filter and neuroplasticity to create effective, lasting change in your classroom and among your students.

The toolkit for each of the 5 steps of Math Therapy includes both *Actions* as well as *Activities*. **Actions** are relatively small adjustments to your current practice and shouldn't take too much time to implement! Most of them will likely mean a slight tweak to your current practice. **Activities** are deeper dives into each step and are more time intensive. I totally get that most of us are strapped for time and feel as if we already have too much content to teach, but if you *can* find the time to squeeze in an Activity or two, it will definitely augment the Math Therapy process. But even if you only have the chance to try the Actions, they will still have a huge impact on helping your students heal their math trauma and build better relationships with math. Start where you are and do what you can—it WILL make meaningful difference!

You'll see that I've included grade-level modifications for the Actions and the Activities. You can also make additional age-appropriate shifts in the way you approach the content. As always, do what works best for your classroom and your students!

Use the legend that follows to decide when you have time to implement the following Actions or Activities!

LEGEND	TIME NEEDED
	0–15 minutes / week These Actions and Activities require little to no implementation time and can be done by making small adjustments to your current practices or simply replacing some of your practices entirely. Some contain components that may take 5 to 15 minutes of class time per week.
	1–3 lessons / term These Activities are more intensive and may require you to dedicate several lessons to implement.
	1–3 lessons / term + some work outside of class In addition to implementation in the classroom, these Activities may require your students to do some work outside of the classroom.

Remember, the *Mythbust* Toolkit will be more and more effective as you start to build a Math Therapy classroom! It's always important to set the stage when you're introducing a new paradigm or way of thinking that will challenge the status quo in math class. As an example, kids have been told that mistakes are bad in math class, and you need to mythbust that! Start a discussion about why mistakes might be useful and challenge students to not only define the word "mistake" but to think of mistakes they've made in their lives that have taught them a valuable lesson. Explain that making mistakes is an integral part of math. You can reference famous mathematicians and their math journeys, or you can talk about how the job of most mathematicians and scientists IS to literally MAKE MISTAKES! Scientists don't come up with the cure for a disease in 1 day—if they did, they'd be out of a job. Similarly, kids have gone through most of their life being told that failure is BAD and that being good at math has literally everything to do with speed and accuracy. WE ARE HERE TO *MYTHBUST* ALL OF THAT! While we can't mythbust a lifetime of messaging in ONE day, we *can* start by methodically putting the pieces together to foster a Math Therapy classroom. This is done in both micro and macro ways, through Actions as well as standalone Activities.

Mythbust Actions

Use the following Actions as a part of your regular classroom practice to mythbust student misconceptions around math!

⏳ Mythbust *Action 1: Offer Opportunities for Reflection*

Why

Mythbust the idea that math is just about getting the right answer and moving on by communicating instead that it is a subject full of opportunities for creativity and growth.

Giving students dedicated time for reflection not only shows them that growth from past experience is a valued part of learning, but it also helps us, as educators, understand our students' math experiences on a deeper level!

When

Include opportunities for reflection as part of every test or assignment! Additionally, you might consider giving students 5 to 10 minutes at the end of each week to reflect on their math experience that week.

How

1. Pop it in! Add a reflection question at the end of any assessment, formative or summative. If you're grading the assessment, be sure you grade these questions as a part of the assessment so that students can see that they are just as valued as "correct answers." Here are some ideas:

   a. *(Reflect on learning process)* Did you notice any particular learning strategies that worked well for you during this test?

b. *(Reflect on challenges)* What was the most challenging question for you, and how did you approach it?

c. *(Reflect on problem-solving)* Describe the strategies you used to solve a particularly complex problem. How did you decide on these strategies?

d. *(Reflect on mistakes)* How can you use spots where you struggled on this test as opportunities for learning and improvement?

e. *(Reflect on progress)* What specific skills or concepts do you feel more confident in after taking this test?

f. *(Reflect on future growth)* What is one area in which you would like to improve your understanding or skills in math? What steps can you take to achieve this improvement, both in and outside of the classroom?

2. Dedicate time! Set aside 10 minutes at the end of each week for reflection. Students can use their Math Therapy journals to answer a prompt of your choosing (see above for some ideas that you can tweak by replacing "this test" with "this week"). For best results, collect these reflections and look them over, providing students with direct feedback and considering an adjustment to pedagogy as needed (see Figures 3.3 and 3.4).

Grade-Level Modifications

- **Grades K–2:** Provide sentence starters and prompt students to fill in the blanks! For example: Doing math makes me feel _____ (happy/sad/mad/proud).

- **Grades 3–5:** Allow students to put their reflections in the form of either a drawing or by jotting down keywords instead of using full sentences.

- **Grades 6–8:** Consider giving students mini-surveys at the end of tests or assignments that ask them to reflect on their feelings on a sliding scale!

- **Grades 9–12:** Give students the option to answer prompts in full sentences or to make point-form notes.

FIGURE 3.3 GRADE 5 END-OF-CHAPTER REFLECTIONS

• I love these end-of-chapter reflections that educator Meggan Dodge uses with her Grade 5 students! Each box has a specific prompt: What are you proud of accomplishing? What is your overall attitude toward math right now? What was one thing you struggled with? How did you overcome that struggle? What would you like to do differently next time? How do you learn best?

SOURCE: Lively Emich, Alexis Emich, & Meggan Dodge

FIGURE 3.4 MORE REFLECTIONS

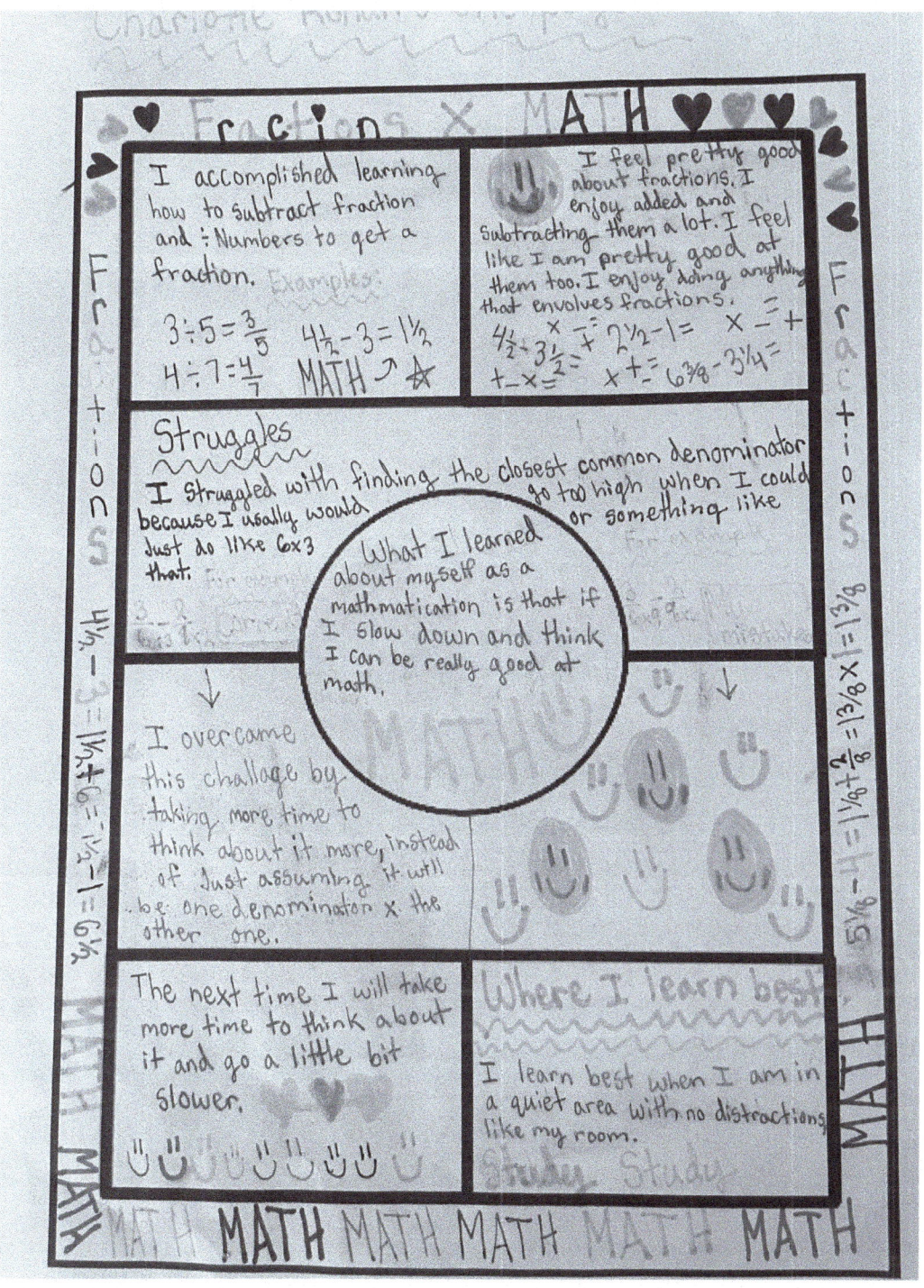

SOURCE: Meggan Dodge

Mythbust *Action 2: Celebrate Effort and Perseverance*

Why

Mythbust the idea that the only thing that matters is getting the right answer.

If we tell students that effort and perseverance are an important part of the learning process, we have to *show* them that we mean what we say! Remember, students pay more attention to what we *do* than what we *say*.

When

Incorporate this strategy daily as a part of your regular classroom practice as well as a part of both your summative and formative assessment practices when possible.

How

1. Adjust your rubrics! Incorporate process, effort, and progress into your rubrics while honoring different ways of knowing. Here are some quick swaps you might consider making:

ORIGINAL RUBRIC	MYTHBUSTED RUBRIC
Use of formulas: Correct application of relevant formulas.	Demonstrates an understanding of multiple problem-solving strategies and employs them effectively, even if the initial approach differs.
Problem-solving: Correctness of the final answer.	Problem-solving process: Demonstrates a thoughtful and systematic approach to problem-solving, including intermediate steps, regardless of the final answer.
Communication: Clarity of the solution presentation.	Communication and reflection: Effectively communicates the solution and reflects on the learning process, including challenges faced and strategies employed.
Add a row for creativity if your current rubric doesn't have one!	Creative problem-solving: Encourages innovative approaches and creative thinking in problem-solving, recognizing that there may be multiple valid ways to approach a problem.

2. Adjust the little things! When you're dishing out compliments, find opportunities to congratulate students on their hard work, effort, perseverance, creativity, and process as opposed to just the correct solution. Hearing this revised narrative over and over, in different ways (e.g., in a rubric, verbally from you, etc.) helps students internalize the idea that growth is valued as an important part of the learning process.

3. Watch out for self-talk! The way you talk about yourself can indirectly impact students as well. Try to avoid judgment statements like, "I'm just the *type of person* who doesn't like to take risks," or "*No matter how hard I try*, I just can't sing on key." While these statements may seem benign, they emphasize the false narrative that our abilities are fixed instead of in flux. Instead, you might say something like, "I just don't like taking risks," or "I can't sing on key." Notice the latter personalizes the statement while the former implies that there is a certain "type" of person who is capable of a certain skill and another group that simply isn't!

Grade-Level Modifications

- **Grades K–2:** Encourage discussion around the value of progress and effort as well as what progress and effort might look like! Reinforce the message with visual reminders such as classroom posters or info charts.

- **Grades 3–5:** Little stickers go a long way! Consider handing out stickers for things like hard work, resilience, helping peers, collaboration, and respectful communication.

- **Grades 6–8:** Encourage students to gently monitor one another's self-talk! If a student is being mean to themselves, prompt classmates to pitch in to help them reframe and reword their thought to make it a little kinder!

- **Grades 9–12:** Make time for discussion around what a rubric should include and why. Allow students to critique your rubrics and be willing to compromise and make adjustments accordingly!

 Mythbust *Action 3: Highlight Awesome Mistakes*

Why

Bust the myth that "failure" and "mistake-making" are bad words in math class and celebrate errors as stepping stones to understanding and improvement.

This Action is all about shining a spotlight on how mistakes are a valuable part of the learning process!

When

Incorporate this strategy daily—or, at the least, weekly—as a regular part of your regular classroom practice.

How

1. Let your students know that you're going to be on the lookout for awesome mistakes to showcase weekly. Awesome mistakes are mistakes that provide opportunities for learning and growth!

2. Play detective! Awesome mistakes can be found anywhere. If you use Liljedahl's (2021) *Thinking Classroom* methodology, you likely have your students solving math problems on a vertical non-permanent surface. Liljedahl suggests walking around the room and circling awesome mistakes in red marker and having students explain their thinking. This is one way to spot and showcase awesome mistakes. If you don't have access to enough non-permanent vertical surfaces for all students to work on simultaneously, have students begin by working at their desks, preferably in groups of threes. This allows students to make their mistakes *together*, ultimately making them more willing to share it with the class, as it isn't just a reflection of their own thinking, but the thinking of their entire group. As you walk from desk to desk, ask students to explain their thinking and select your favorite mistakes for showcasing. One or two mistakes per lesson are enough to demonstrate how valuable mistakes are and to reinforce a classroom culture where mistake-making is embraced (see Figure 3.5).

FIGURE 3.5   EXAMPLES OF FAVORITE MISTAKES

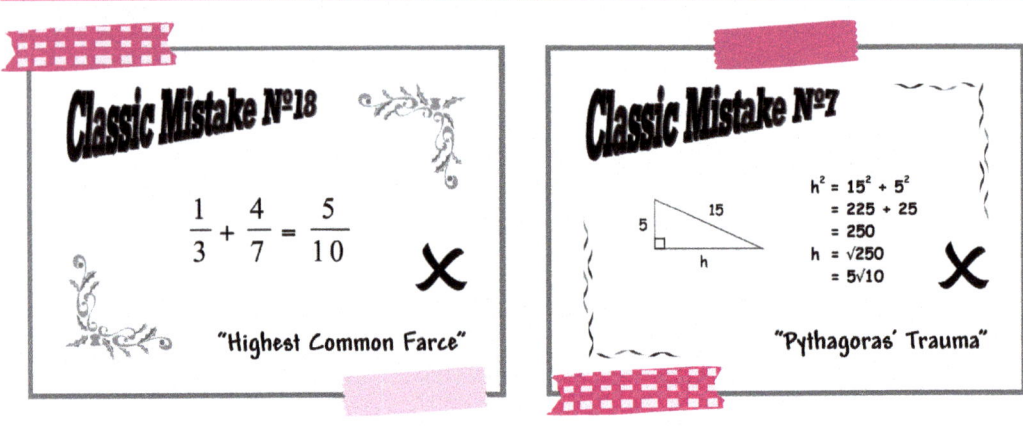

• Educator Richard Catterall has plastered his classroom walls with his favorite mistakes, created by Nevil Hopley! You can find out more and download your own posters at https://classicmistake.nhost.uk/.

SOURCE: Nevil Hopley

HOT TIP

Set parameters around what an awesome mistake *is*. Awesome mistakes are not mistakes made on purpose for the sake of having something to showcase. Those are *meh* mistakes. Awesome mistakes are mistakes that organically come up and that we have something to learn from—that's what we're looking for here!

3. Show and tell! This is where you really get to mythbust the idea that mistakes are bad by showing your students that you care about their process and progress just as much as you care about "getting the right answer." When you spot a mistake you love, tell students that you think they've made an AWESOME mistake. You might say something like, "OMG, so many of my students do this exact same thing—I would love for you to explain your thinking here so that we can all figure out how to avoid this common pitfall!" or "I LOVE the creativity of this mistake—I've actually never seen anyone do anything quite like this, and I would love for you to explain your thought process!" Hype your students up! As students present their work in front of the class, you are not only celebrating their willingness to grow and learn from their own mistake, but you are showing the entire class that making mistakes isn't something to shy away from. This emphasizes the importance of having a growth mindset, and it empowers kids to raise their hand or take a stab at that test question they're not sure about. It encourages your students to take the risks needed to learn new math concepts and try new techniques, and ultimately, this improves not only their relationship with math—but their performance as well.

Grade-Level Modifications

- **Grades K–2:** Grab a detective cap and a magnifying glass and get into the role of playing detective when you're on the lookout for spicy mistakes!

- **Grades 3–5:** To get students comfy with the idea of making mistakes, make your own mistakes up on the board and invite kids to point them out! This allows them to work together and to focus on *your* mistakes instead of their own, which at first can seem more fun and less daunting—everyone loves pointing out when the teacher is wrong!

- **Grades 6–8:** When mistake show-and-tell is happening, allow students in the class to ask the presenter questions as though this is a press conference! Make it fun and have the presenter play the role of a celeb being interviewed!

- **Grades 9–12:** Circle great mistakes on student tests. When you take up the test for review, ask selected students if you can use their amazing mistakes as teaching tools—this enables students to feel successful on a test even if they didn't get the answer or grade they were hoping for!

Mythbust Activities

Use the following Activities to engage students in mythbusting misconceptions around math on a deeper level!

 Mythbust *Activity 1: My Math Therapy Journal*

 This free download is available online at **maththerapy.com**

Why

The journal will create a dedicated space for each student to record their Math Therapy journey, enabling them to reflect on their growth, in real time, over the course of the year.

When

Remember way back when you started this book and I mentioned that I would be talking about how to create a Math Therapy journal for you and your students? Well, here we are! Incorporate this strategy weekly—as a regular part of your classroom practice. If you have the time for it, reserving 5 to 7 minutes at the end of each lesson for journaling is the absolute best-case scenario, but don't sweat it if that's just too much!

How

1. Dedicate space! Start this Activity by asking students to pick a dedicated space for their Math Therapy journal. I have included a downloadable template for Math Therapy journal pages that your students can use for this Activity as well as many of the other Activities!

2. Prompts! We rarely ask our students to actually reflect or *feel* in math class, and both of those things are crucial to building a better relationship with math. Because most students aren't used to journaling, providing prompts to help them explore their relationship with math is a great way to get them going. There are SO many potential journal prompts, and you can totally feel free to get creative and make up your own—you know your students best! In terms of timing, I find it works best when you carve out 5 to 10 minutes at the beginning of class, give students a prompt, and then set a timer. Tell them to just start writing—anything—on the page, and before they know it, time will be up!

Here are some prompts I love:

- Describe a recent math challenge. How did you approach it, and what did you learn?

- Think about a recent math mistake. How did you feel, and what steps can you take to avoid similar mistakes?

- Pick a math concept you found hard initially. Describe your progress and the steps that helped you improve.

- Recall a time when you collaborated on a math project. How did it impact your understanding?

- Reflect on positive math feedback. How did it affect your future approach?

- Think about a time when you persevered through a tough problem in math. What motivated you, and what did you learn?

- Describe a situation where you used a different strategy in math. How did it change your understanding?

- Imagine mentoring a younger student in math. What advice would you give about developing a positive attitude and facing challenges?

3. Populate the pages! Throughout the rest of this book, you will be able to download templates for many of the Activities that I will be sharing—you can use those to populate your students' Math Therapy journals! There are also a ton of great resources out there, so if you see a growth mindset activity that you like, feel free to throw that in as well!

4. Peek! We want the Math Therapy journal to be a safe space for students to reflect on their thoughts, and sometimes the knowledge that they have to turn it in to an adult for review may hinder that. That being said, one way to combat this is to make sure your students know that they are *absolutely not being graded* on the content of their journals and that you're only sneaking a peek to help them build better relationships with math. If you feel that your classroom culture allows it, you can ask students if they might want to share their responses to some of the provided prompts. Over time, as more students share and as your classroom becomes a Math Therapy classroom, you will find that students not only love journal time—but love sharing their experiences with one another and feeling validated for not being the only ones who have complicated feelings about math!

- **Grades K–2:** Instead of asking students to write out complete sentences, provide sentence starters and options so they can fill in the blanks. For example: "Today's math lesson made me feel _____ (sad/happy/angry/proud)."

- **Grades 3–5:** Simplify the prompts so that the language meets the requirements of your grade level and give students ample space to write. You might also allow them to respond to prompts by drawing as well as writing.

- **Grades 6–8:** In addition to adjusting the prompts to meet the needs of your grade band, consider beginning this Activity with a mindfulness exercise so that students are able to get in the zone! An easy way to do this is to start with a mindfulness minute: Set a timer for 1 minute and ask everyone to close their eyes, sit upright with their feet planted on the floor, and focus on their breath. You'll be shocked at how quickly this transforms the energy of your classroom!

- **Grades 9–12:** The prompts provided are targeted to this grade level, but you know your class best! If needed, simplify the prompts and give your students dedicated time to write and reflect. Allow them to listen to music, put in ear buds, or negotiate what might be needed for them to "get in the zone!"

 Mythbust *Activity 2: I Used to Think, Now I Think . . .*

Why

Mythbust misconceptions about fixed mindset by showing students that they are living proof that the brain can grow and change.

> I used to think I couldn't understand negative numbers.
>
> Now I think they totally make sense!

We all know that you can tell someone something over and over again, but until they experience it *for themselves*, they're unlikely to believe you! That's why this Activity is all about allowing students to come to their own conclusion that growth mindset is the real deal—by looking at factual evidence from their own lives!

When

The bulk of this Activity can take place over one lesson. Reserve 5 to 10 minutes each week for students to add to their growing list of things they used to think they couldn't do but now can!

How

1. Build up curiosity! Start this Activity by asking students to consider whether or not they believe that they can personally learn new skills. Prompt them to discuss whether they think that *all* skills can be learned; this isn't just about math skills—it's about ALL skills. Are there certain skills that are easier to learn than others? Are there certain skills that simply can't be learned? Feel free to allow students to discuss in pairs and to share their thoughts with the class.

2. Mythbust it! You will assign students the task of making a list of three things they *used to think* they couldn't do that *now* they can do. Give them some examples! I personally used to think that I would never be able to raise my left eyebrow, but after MONTHS of practice, I can finally do it! I used to think I could never learn to surf, but I did eventually learn how to get up on a surfboard. I used to think that I would never have the patience and discipline to write a book, yet here we are . . . and so on! If you have decided to implement a growth mindset journal (Activity 1), you can have your students start this list in there. If not, you can have them do the Activity in their math notebook or on a piece of poster paper that you will put up on the wall. Whichever you choose, make sure your students have a dedicated space to make their list that they can easily find each week—you are going to build on this Activity weekly throughout the year so that any ideas of fixed mindset get busted consistently and repeatedly!

3. Let's hear it! Once your students have their list of three things, you can have students share with the class. Engage in a discussion about what it took for them to get from the "I used to think (I couldn't) do . . ." part to the "Now I think (I can) . . ." part. This is an opportunity to discuss the *behave* part of growth mindset and to crowdsource concrete actions and steps students can take in your class to progress in their math journey. Hearing from other classmates will not only help students feel like these actions and steps are within reach and actually work, but it will generate additional buzz to the mythbusting discussion!

4. Rinse and repeat! This is the best part. Each week, ask students to add on a new "I used to think, now I think . . ." You can ask students to add on one that has nothing to do with math (e.g., I used to think I couldn't sing in public, but now I think I can because I did karaoke last week and didn't die!) and one that is math-specific (e.g., I used to think I could never add fractions, but now I think I actually can because something finally clicked last week!). Again, if these lists are somewhere public, like on your classroom walls, students can add on to those visible lists, which will pump up the rest of the class—trust me! If those lists are private or in a journal, equally great! Just make sure students are able to access those lists weekly so that they can add to them and see how far they've come. Center discussions around these lists monthly in the same way you did in Part 3 of this Activity!

Grade-Level Modifications

- **Grades K–2:** Create posters for your classroom walls so that students can always access visual representations of "I used to think, now I think" statements.

- **Grades 3–5:** Start with a group discussion so that students can see the pattern of the Activity. Share some of your own "I used to think, now I think" statements. You might start by having kids choose ONE of their "I used to think, now I think" statements to turn into a classroom poster. Each student can embellish their poster with glitter, doodles, and stickers so that the room is covered with student work that mythbusts misconceptions about abilities being fixed rather than fluid.

- **Grades 6–8:** Give students some concrete examples of your own "I used to think, now I think" stories so that they can understand how the pattern of the Activity works and how diverse the "I used to think, now I think" statements can be. Make sure you give them several very different examples, either from your own life or the life of a culturally relevant fictional or nonfictional character (e.g., T'Challa, Wonder Woman, Taylor Swift, SpongeBob, or whatever kids are into these days).

- **Grades 9–12:** Students might feel self-conscious in revealing what they once thought they couldn't do that now they can do. Build trust by sharing your own "I used to think, now I think" statements and encourage students to share their own. If necessary, you can have them do this Activity in their journal and ensure them that no one will see their list but you.

Mythbust *Activity 3: My Math Superpower*

Why

Bust the myth that being "good" at math has to do solely with speed, ease, or correctness, and showcase all of the ways in which math happens!

The way we define *math* is SO narrow that it's no wonder many students feel unworthy of being labeled "good at math." This Activity will not only help kids define math in a way that feels meaningful to THEM, but it will equip each student with their very own unique math superpower so that they can see that they have something valuable to contribute in math class! I have personally done this ENTIRE Activity with students over the course of a 1-hour lesson, but other teachers have told me that this Activity works best by dedicating an entire lesson to busting math myths and a subsequent lesson to the creation of each student's math superpower.

When

This Activity can take place over the course of one or two lessons. The earlier in the year you can set students up to see that math encompasses a broad set of skills and that they all have something unique to contribute, the better!

How

1. Bust away! Start with a discussion of what it means to be "good at math." Write student answers on the board and get a list going, leaving space underneath each "myth" so that you can bust them later. Once you have a list of math myths on the board, let the busting process begin! Ask students if they can prove that each of these myths are true or false. This is also a great opportunity for discussion on what "proving" something even means—in math class—and beyond! The following screenshots are from a session I did with Grade 8 students over Zoom. We came up with the myths together, then discussed examples that proved them wrong. Then, we wrote a statement to replace the myth with a new fact!

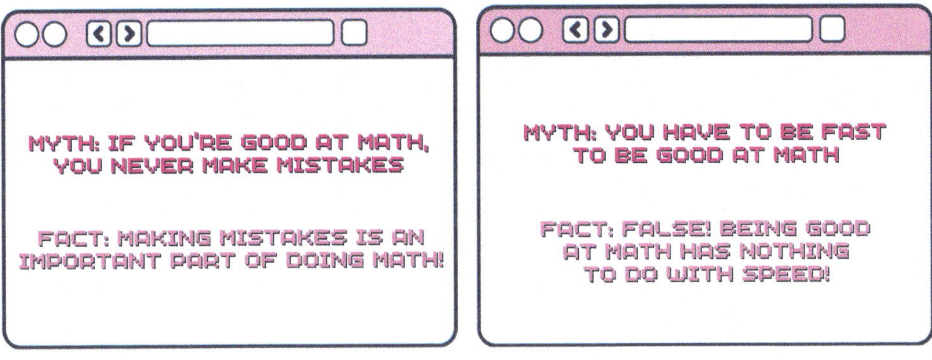

2. Broaden the definition of math! Now that you've busted some myths about being GOOD at math, it's time to talk about what math even IS and to broaden the scope of what a "math skill" might look like. Generate a word map with your students as a group so that everyone can add to the conversation and rich discussion can ensue.

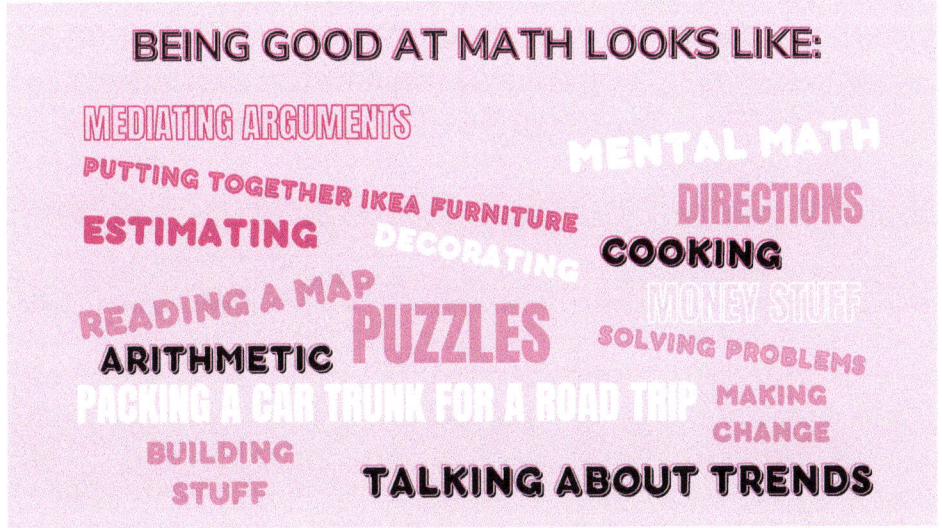

3. Superpower time! This is the most exciting part. Now that your students have redefined what being "good" at math really means and have broadened the scope of what math is, it's time to find each of their unique superpowers. You can do this individually, in pairs, or as a group. Figure 3.6 shows an example of what I used as a template to do this Activity with that same group of Grade 8 students over Zoom. (The template is available free on the Math Therapy website.) One by one, they came up to the camera and asked to be assigned their math superpower. The rest of the class pitched in, and it was so fun! There are three parts to this Activity:

a. Ask each student what they love to do and already feel good at (e.g., soccer, singing, hanging out with friends).

b. Now, ask what skills they use when they are doing that thing (e.g., teamwork, vocal warm-ups, solving friend-group problems). Other students can start to pitch in here if they're having trouble coming up with things!

c. Finally, ask how they feel they might transfer those skills *to* math (e.g., collaborating, doing math warm-ups, coming up with creative solutions to problems). This is a great time for other students to pitch in! Now, they are ready to be assigned their math superpower.

FIGURE 3.6   EXAMPLES OF MATH SUPERPOWER CERTIFICATES

4. Frame it! Download my printable math superpower certificate and assign each student a superpower! When I did this with a Grade 12 calculus class, the kids loved coming up with the superpower tagline together. One student said he loved creative writing and was really good at deciphering what stories and songs were about, so his math superpower became "the poet of the problem." Another said that she loved playing sports and doing drills to get stronger, so her math superpower was "I'm not afraid to get down and dirty!" You can get as creative as you like with it, but once it's on paper, encourage kids to reference that skill throughout the year when in math class. You can revisit this Activity, come up with additional superpowers, and even stick these certificates on the wall so that your students can remember all the myths about math you mythbusted together and appreciate the diversity of thinking that makes mathematics for everyone!

Grade-Level Modifications

- **Grades K–2:** Get into the drama of it all by having students dress up as superheroes that match their math superpowers!

- **Grades 3–5:** If it's too time intensive to have the whole class work together to come up with a superpower for each student, try having kids work in pairs to help each other come up with their superpowers!

- **Grades 6–8:** Sharing is caring! Have kids work in pairs to help each other discover their superpower! Have each pair present their superpower to the class—have one student present the OTHER'S superpower and vice versa. This helps foster a Math Therapy classroom where students are a part of helping one another feel good about math!

- **Grades 9–12:** Consider having students work in groups of three to help each other come up with their individual superpower. Encourage them to get creative with how they phrase the superpower for presentation on the actual certificate and present them with certificate options (I have included several templates on the Math Therapy website for them to choose from)! Each group can then present their superpower to the class—have one student present the OTHER'S superpower and vice versa. This helps foster a Math Therapy classroom where students are excited about helping one another feel successful!

Mythbust *Activity 4: Seeing Stars*

Why

Mythbust the idea that the way we think is fixed and can't change.

This Activity shows students how malleable the mind is and that what you focus on really does grow, including our thoughts about our own math abilities! A bonus of this Activity is that students will open their eyes to spotting shapes and patterns, a math skill that often goes uncelebrated!

When

This Activity can be done over the course of 1 day (younger grade bands) or two lessons *plus* a little homework (older grade bands)!

How

1. Explain! Begin by introducing your students to the concept of the RAS filter. You can get as science-y as you want with them, but the point here is to engage in a discussion about the nature of our thoughts as they relate to math. We want to mythbust the idea that our thoughts are fixed and guide our students into realizing that they have some power over how they think about math and that, by tapping into that power, they CAN begin to develop more positive thoughts about math. Some questions you might use to engage them include:

   - What are thoughts?

   - Are thoughts the same as facts? Why?

   - Do we have control over our thoughts?

   - Can we train our brains to focus on positive thoughts instead of negative thoughts?

   - What does "what you focus on grows" mean? Do you believe that the saying is true? Why?

2. Set up the challenge! Explain that participants will be actively looking for star shapes in their environment. You can even take a moment to talk about what defines a star. How many points can a star have? What are the characteristics of a star? Work together to lay out some parameters to get your students invested. Explain that these star shapes can be found in various forms, such as in nature, objects, patterns, or even created by an arrangement of items; there are no rules—anything goes!

3. Journal it! Instruct your students to open their Math Therapy journals and dedicate a page to this Activity. They should create a simple chart with columns for Date, Time, Location, and a Description (brief) of the star they spot.

4. Ready, set, go! Tell your students they have 24 hours to complete this Activity! Make sure to emphasize that the key here is to be open-minded and focused on their mission: Stars can be found anywhere, so students should look closely at their surroundings!

5. Show and tell! After 24 hours have elapsed, have everyone bring their results to the table and discuss their findings. Here, you can have fun with it by creating awards for "the most creative star" or "the most unlikely location" for a star to be found. Remember, this is an opportunity for rich discussion about patterns, shapes, and most importantly—mindset! Some great discussion questions include:

   - What surprised you most about this Activity? Least?

   - Did you see more stars than you thought you were going to see? Explain!

   - What does this Activity suggest about the way our minds work when it comes to focusing on positive or negative things?

   - Now, what do you think about the saying "What you focus on grows"?

   - We tried this Activity with stars, but what else do you want to try to train your brain to focus on?

   - Do you think you could use this same technique to think more positive thoughts about math?

6. You might give students 5 to 10 minutes to record their thoughts and feelings in their journal after your juicy class discussion; make sure you give them a direct prompt to respond to that ties the Activity back to the *Mythbusting* mindset by recognizing their ability to build a better relationship with math by choosing where and how to focus their thoughts (see the prompt ideas presented earlier)!

Grade-Level Modifications

- **Grades K–2:** Bring in manipulatives to kickstart the activity and have students create stars out of construction paper, blocks, tiles, or anything else you have lying around.

- **Grades 3–5:** Instead of making this a 24-hour Activity, allow all the star-spotting to take place within the school day. Assign this Activity in the morning and make sure you take moments throughout the day to point out different stars that appear (maybe the sunlight creates a star-shape on the ceiling or students are seated in a star shape or star stickers are spotted on the front of a book). Encourage students to do the same. The following day, engage in the discussion part of the Activity. When you lead up to the Activity, instead of going into depth about the RAS, have a gentle conversation about the nature of thoughts and adjust the provided prompts accordingly!

- **Grades 6–8:** Once you assign this Activity, make sure you take moments throughout your lesson to point out different stars that appear (a student doodle, a star-shaped twinkle in a light bulb, a shape in a textbook, etc.). Encourage students to do the same so that they are primed for what to look for after the bell rings! When you lead up to the Activity, instead of going into depth about the RAS, have a gentle conversation about the nature of thoughts and adjust the provided prompts accordingly!

- **Grades 9–12:** There is *a lot* of room for very interesting discussion around the nature of thoughts with this grade band! Consider splitting this Activity into two lessons: the first being the nature of thoughts and the RAS and the second being the introduction to the star-spotting Activity.

MYTHBUST TOOLKIT: SIMPLE SWAPS

Want to take your mythbusting to the next level? Use the following simple swaps to effectively sub out some of your current practices for those that will help you *Mythbust* your way to a Math Therapy classroom!

SWAP THIS	FOR THIS
Erasing mistakes on the board (your own or your students'!)	Leaving mistakes up for everyone to see
Fixed language ("I'm not good at this")	Growth mindset language ("I can improve with practice")
Fixed praise ("You're so smart")	Praise for effort ("I appreciate your hard work")
Avoiding challenges	Providing opportunities for ALL students to be challenged
Strict time constraints	Providing time for mastery

MYTHBUST TOOLKIT: THEY SAY, YOU SAY!

One of the most common questions I get is, "But how do I respond when my students say . . . ?!" I totally feel you. That's why I've created a table of common questions, along with potential responses you might want to try, as a part of every single toolkit in this book! Here's your first one, and I want to caveat it by suggesting that, whenever you can, prompt your students to think of examples where your response can be illustrated by a real example *from their own lives*. Trust me, this is so much more effective than you being, like, "Just believe that what I say is true without any actual proof!" The more you can prompt your students to draw on their own lived experience to bring your responses to life, the more impactful your words will be!

WHEN THEY SAY . . .	YOU SAY . . .
"It takes so much longer for me to get it than anyone else!"	"That's totally fine! It took a long time for me to learn how to _____ (fill in the blank), but that doesn't make me bad at it!"
"But my brain just doesn't work that way."	"It may feel that way now, but the more you train your brain to think in different ways, the better it gets at it! Name one of the ways you think your brain 'works' now—can you think of how you trained your brain to work that way?"
"No one in my family is good at math."	"Good thing math isn't genetic!"
"I'm just not a math person."	"What's your definition of a math person? What would it look like if you just did math on your own terms?"
"I don't get it!"	" . . . yet!" (pause to allow them to roll their eyes) "Just because you don't understand something this second doesn't mean you never will!"

 ## *MYTHBUST* TOOLKIT: EXPANSION PACK

I wish I had a million more pages to include all of the *Mythbust* Activities that I've used over the years, but hey, that's what my website is for! The following are more of my fave Activities. You can find templates and instructions for most of these Activities by heading to the Math Therapy website.

FIGURE 3.7 MY STUDIO'S FAILURE WALL

1. **Make Your Own Failure Wall:** Mythbust the idea that failure is a bad thing by showcasing both famous failures and student failures. Figure 3.7 shows a photo of the Failure Wall at my tutoring studio. We have selected celebrities we love and included a card with their failure story below each picture, like an art exhibit. Then, we created a bulletin board where students consistently add on cards with mistakes or failures they've had on the front—and on the back, the lessons they learned from them!

HOT TIP

Set some parameters around *who* might be appropriate to put on the failure wall. I say this because a teacher emailed the week before I wrote this chapter to say that she had assigned this failure wall assignment through a substitute teacher, so she had given her students no parameters or context. She came back the next day and one of her students had chosen Osama Bin Laden to add to the failure wall . . . I'll leave it at that!

2. **Mathfirmation Deck:** Mythbust the belief that our thoughts can't change by creating positive affirmations that help augment a growth mindset. Remember, what you focus on grows! Have students create their own math-related affirmations on index cards or get crafty with it and spend a day using glitter, stickers, and other supplies to create unique Mathfirmations. Encourage them to personalize the cards and keep them visible during class as a positive reminder and feel free to download my template for added inspo! Some of my fave Mathfirmations include *I can do hard things* and *Mistakes help me learn and grow*.

3. **Guest Speaker Series:** Mythbust what math even IS by showing your students that math means so much more than just speed and equations! Invite folks from math-related fields to tell students about their math journey and explain what kind of math skills they use day to day! When you're searching for speakers, consider perusing speaker bureau websites; many of them have entire pages devoted to diverse STEM speakers. Another great way to find speakers is by searching popular math education conference sites, such as NCTM, CAMT, and GYTO, to find out who is speaking on what topic. Finally, many of the cool folks I know who speak about math and math-related topics can be found on social media, so ask your teacher friends what mathfluencers are trending and use that as a starting point!

4. **Failure Certificate Exercise:** Celebrate student failures by helping them create their very own failure certificate! Download one of the templates provided on the Math Therapy website and hand them out to your students. Ask them to think of a big failure they learned a lesson from. Then, have them chat with a partner to talk it out before they write their big failure on their certificate, along with the lesson it taught them. If possible, display these certificates on your classroom wall (think: gallery vibes!) so that your students can be reminded that failure is nothing to be ashamed of but something to celebrate as part of the path to success! *Warning*: Set some parameters around this Activity or you *may* get some NSFW responses!

5. **DIY Failure Quotes:** Have students come up with their favorite failure quote and engage them in a "failure quote show and tell" session! There are millions of failure quotes online, so have students take a day to find their favorite one and to write a statement about why they chose it and what it means to them. The following day, have them present to the class. This is a great opportunity for discussion about the many lessons that can be learned from failure and the many approaches we can take toward it!

6. **What Does a Mathematician Look Like?** Mythbust the idea that there is only one way to be a mathematician . . . and the idea that that one way involves looking like Albert Einstein! This is a simple, fun Activity where ALL you do is give kids a simple task: You tell them to draw a mathematician! That's it! Give students 10 minutes to draw their mathematician, then tape all the drawings to the wall, ideally close to each other so students can remark on their similarities and differences. Prompt students to talk about what they notice and wonder and ask students to explain why their mathematician looks the way it does. This is a great opportunity for discussion around stereotypes, media messaging, and representation!

STEP 1 TOOLKIT
TO GO
mythbust

mythbust
ACTIONS

1. OFFER OPPORTUNITIES FOR REFLECTION

Add reflection questions to each assessment and set aside time for reflection prompts.

2. CELEBRATE EFFORT AND PERSEVERANCE

Incorporate "progress" metrics into rubrics, get creative with compliments, and watch your self-talk!

3. HIGHLIGHT AWESOME MISTAKES

Celebrate mistakes by complimenting and showcasing your students' mistakes!

mythbust
ACTIVITIES

1. MY MATH THERAPY JOURNAL

Create a dedicated space for students to record their Math Therapy journey.

2. I USED TO THINK, NOW I THINK

Make a list of things you used to think you couldn't do, that now you can!

3. MY MATH SUPERPOWER

Help students align the math-adjacent skills they're building with the real-life skills they already have!

mythbust
EXPANSION PACK

DIY FAILURE WALL ✱ MATHFIRMATION DECK ✱ GUEST SPEAKER SERIES ✱ FAILURE CERTIFICATES ✱ DIY FAILURE QUOTES ✱ WHAT DOES A MATHEMATICIAN LOOK LIKE?

BEFORE WE MOVE ON

Before we move on to the next chapter, it's important to celebrate what you've learned so far and to take a moment to really reflect and let it all sink in. To help you out, I have a few final treats for you to pack up and take with you!

Treat Yourself

It's time to use what you've learned in this chapter to treat yourself! It's amazing what a few moments of mindfulness can do to replenish and reenergize you, so I want you to pick a mindful activity to treat yourself to. Take a candlelit bubble bath with some yummy bath salts; go for a quiet walk in nature; or literally just lie on the floor with all the lights off, set a timer for 5 minutes, and just listen to the sounds of your breath. Whatever you choose to do, take a few moments to quiet your mind and be still—be here . . . now— you've come so far!

ASK YOURSELF

1. How did this chapter bust some of your own myths about mindset, failure, and being "good" at math?

2. Are the ideas presented in this chapter different from your current educational philosophy? If so, how?

3. Which of your students do you think will benefit most from the Actions and Activities suggested in this chapter? Which do you think won't benefit as much? Why?

4. Which Action or Activity are you most excited to try with your students?

5. What is something in this chapter that makes you go, "aHA, this is totally going to make a difference!" Why?

6. What is something in this chapter that makes you go, "That will NEVER work with my students!" Why?

7. What are some of the challenges you may face when trying to implement the suggestions in this chapter? How might you prepare for these challenges ahead of time?

CHAPTER 4

MATH THERAPY STEP 2

Moderate Math Trauma

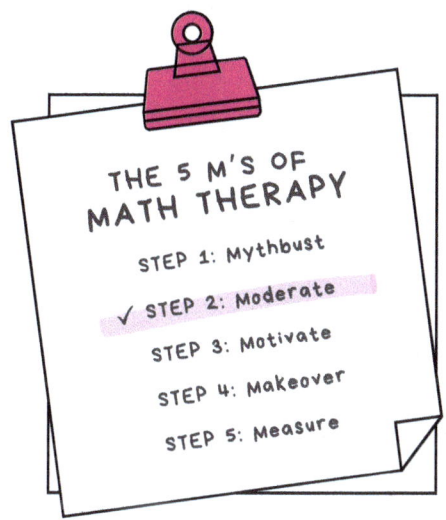

THE 5 M'S OF
MATH THERAPY

STEP 1: Mythbust

✓ STEP 2: Moderate

STEP 3: Motivate

STEP 4: Makeover

STEP 5: Measure

In this chapter, we get to:

1. Explore more about math trauma and how it affects our students

2. Understand why helping our students heal their math trauma is both an offensive *and* defensive practice

3. Dig into Step 2: *Moderating*

4. Jump into a toolkit of concrete Actions and Activities to bring Step 2 of Math Therapy to life in our classrooms in an immediate and impactful way

I met Jaime in Grade 12 when she came to see me for tutoring. She showed up for her session and immediately proclaimed that she wasn't a "math person." She elaborated:

"I make a lot of mistakes."

"Even though I try, I didn't get good grades."

"My friends always tell me that I'm more of an artist, so it makes sense that math isn't my thing and that all I need to do is pass and stop stressing about doing well."

After thanking Jaime for being so vulnerable and digging deep, I shared that, even though it might seem benign and that I'm sure her friends are well-intentioned, that repeatedly telling someone that they shouldn't worry about math because they have "other strengths" inadvertently sends the message that there's no point in them trying to get better at math because it's likely not possible. Jamie had never thought about it that way before and agreed that, actually, she may have gotten that idea along the way without even realizing it! We chatted a bit more, and then she said something that piqued my interest: "Oh, I forgot to tell you, something really exciting happened a few months ago! I was FINALLY diagnosed with ADHD!" Jaime laughed, continuing with, "Now I know why I can't focus on literally anything, especially math!"

I pondered out loud, "OMG, do you think that if you had known that earlier that you might not have developed this idea that you just CAN'T DO math?!"

"Ha," she responded, "maybe!" I was thrilled. Why? Because I was about to blow Jaime's mind.

"Jaime, do you realize how traumatizing it can be to think that there's something wrong with you, only to be told years later that you actually have a legitimate, diagnosable condition, which wasn't just overlooked and ignored, but which comes with a set of resources that you didn't have access to until MONTHS ago?!"

Her eyes widened as she realized the magnitude of this revelation. An undiagnosed learning difference certainly isn't a component of every case of math trauma, but it certainly can be a contributing factor. We were plowing our way through Step 2, and there was no turning back!

In this ONE conversation, before we had even done ANY math, both Jaime and I had already figured out that she had two potential sources of math trauma that had led her to believe she just couldn't do math:

1. *Well-meaning friends who repeatedly told her that math wasn't her strength and that artists aren't typically good at math*

2. *An undiagnosed learning difference that made her feel as if she couldn't do math, no matter how hard she tried*

Jaime and I talked about how both her friends' well-intentioned words, coupled with her late stage ADHD diagnosis, may have contributed to underlying math trauma that prevented her from feeling capable of doing math and of taking the steps she needed to get there. She felt validated for those feelings, and she acknowledged that that trauma was not fixed, nor did it define her. It was simply an obstacle to be unpacked and understood along her math journey, which, as it turned out, was just beginning.

I suggested she keep an open mind as we started our lesson and guaranteed her that she was going to learn AT LEAST SOME MATH over the next hour with me, proving that she WAS capable of improving her relationship with math—something she had decided was a lost cause before she even walked in the door. I told her that now that I knew she had ADHD, we were going to take frequent breaks, minimize distractions, and kick off with a breathing exercise to get us into the zone.

We had an amazing lesson.

At the end of our hour, she didn't understand every single thing, but we made a list of five things she knew that she hadn't known when she walked through the door (throwback to Step 1 Mythbust *Action: I used to think, now I think!). Her math trauma hadn't disappeared, but her understanding of it had, allowing her to remain open-minded to her Math Therapy journey.*

 **IN THE MOMENT**

Imagine you needed glasses, and you just didn't go to the optometrist. You would legit not be able to SEE clearly! It's the same thing that happens when we don't provide opportunities for our students to identify their learning differences. Normalizing learning differences and letting your students know you're there to support them will make a HUGE difference. Do you have any students in your classroom with learning differences? Make sure they feel supported and that they are receiving all of the accommodations available to them. Some of my students have expressed that they're too embarrassed to tell their teachers that they've been diagnosed with a learning difference. Let your class know that learning differences are totally normal and that the more we all know about ourselves as learners, the more support we're open to receiving.

Jaime would go on to do more than pass. We spoke to her teacher about her diagnosis, and she was given an accommodation so that she could have more time to write her tests, which helped tremendously. She got so excited and comfortable with math that she ended up tutoring other students in her class and applying to university programs she had never thought she would get into before. She is now an adult who loves to loudly proclaim, "I used to THINK that I wasn't a math person . . . but not anymore!"

Note that in addition to helping Jaime unpack her trauma, I also helped her understand that math trauma can lead to feelings of inadequacy when it comes to math. I also made sure she felt validated for the complicated feelings she has carried with her, and I let her know that acknowledging where those feelings come from is often the first step toward healing! It's all of it: the unpacking, the validating, and the healing of math trauma that make Step 2 so powerful, so let's get into it!

STEP 2: *MODERATE*!

Next step: Moderating math trauma!

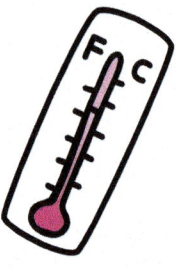

MODERATE

The second step of Math Therapy empowers students to moderate their complicated relationships with math so that they can finally begin to heal and move forward! In this step, I'll provide a toolkit packed with Actions and Activities that you will use to create an environment in which students feel comfortable exploring their—often complex—feelings about math. You will learn how to guide students with math trauma through the healing process, how to prevent retraumatizing those students, and how to avoid causing math trauma in those students who haven't yet had negative experiences with math! This step is all about facing math trauma head on so that the healing process can truly begin!

Why It Works: Once you've invested some time on mythbusting student misconceptions, it's time to introduce the concept of math trauma. The step of moderating math trauma catalyzes students' feelings of validation and helps them to understand that their complex feelings around math aren't necessarily a reflection of their *ability*, but of the complex factors that surround them!

HOT TIP

Advocate for yourself! Don't ever let someone tell you that a problem you're having has no solution. Whether it's inside or outside of the classroom, *every* problem is worth identifying and attempting to solve. Even if you don't find a solution right away, exploring different pathways and talking to others about your situation will, at the very least, lead to different lenses for you to view your problem with!

How It Works: What does it even mean to "moderate" math trauma? The truth is, I struggled to find the perfect word for Step 2 and chose this one because it encompasses the delicate complexity of helping students unpack math trauma! This step works by introducing students to the emotional component of learning math. In this step, our goal is to help our students answer a very important question that likely no one before has given them permission to ask:

Why DO I feel this way about math?

The word "moderate" has several meanings that are relevant to Step 2:

1. **To watch out for:** In Step 2, we both learn how to spot areas where students may have math trauma AND help our students spot those areas themselves!

2. **To make less intense:** In this step, we help our students develop strategies to mitigate negative feelings about math.

3. **To observe and adjust the speed, intensity, or strength of something:** Equipping students to moderate their math trauma, or put it into perspective, means that side effects of math trauma such as anxiety and defeat are much less likely to interfere with their ability to build a better relationship with math!

By simply introducing the idea that there is an emotional component around learning math that EVERYONE experiences, we are letting students know that if they currently don't have a great relationship with math, they're not alone, and THERE IS LIKELY A REASON behind it!

 **IN THE MOMENT**

Which version of "moderate" resonates most with you? Are there any I didn't mention that you can bring to your exploration of math trauma in the classroom?

The Step 2 *Moderate* Toolkit is both reactive *and* proactive. On one hand, it's full of tips and tricks to help you learn how to spot math trauma, respond to math trauma, and, ultimately, to help your students unpack and heal from past math trauma so that they can build a better relationship with math (reactive); on the other hand, it's also full of techniques and strategies to prevent the development of math trauma in your classroom in the first place (proactive).

Reactive

Many of our students are going to enter our classrooms with math trauma. The older the kids are that you teach, the more accurate that statement gets. Remember, math trauma is often not just the experience of a singular event but gets even more embedded with repetition and consistency. So if a Grade 1 female student (let's call her Paris) enters your classroom and has had an experience with a parent who has told them, "Don't worry that you don't understand math as well as your brother does; boys are just better at math,"

that can definitely be traumatic. Paris might enter your classroom believing that she is innately incapable of math; she might lack confidence; she might feel anxious and stressed. Now imagine you have Paris as a student in Grade 9. She has now not only had a parent who has made it clear that, in their mind, boys are better at math, but she's been exposed to a ton of media that reinforces the same message. She's wondering why even though *The Big Bang Theory* is ALL about math, Penny, the "hot blonde neighbor" happens to be the one who's *not* good at math, while Amy, painted as a total nerd, *is* good at math. She's wondering why, even though it's 2024, she's still never seen a movie where the cheerleader character is good at math. She's been flipping through 8 years' worth of math textbooks and has barely ever seen a woman on ANY of those pages. In Grade 5, she had a teacher who was likely well-meaning, but always seemed to pick boys to answer questions instead of the girls in her class. I could go on but . . . you get my point, right?

All of these instances lent themselves to the repetition and consistency of this salient message: *Girls aren't as good at math as boys. You are a girl. You probably suck at math and don't belong here.* How icky does it feel to even read that? Now imagine how it might feel for Paris! Over time, those micro traumas build up, and by the time Paris gets to you in Grade 9, she is totally disengaged. She doesn't necessarily even seem anxious or stressed out, she doesn't seem to care, she doesn't seem to try, and she always seems like she's not paying attention. Can you blame her?

> **Remember, math trauma is often not just the experience of a singular event but gets even more embedded with repetition and consistency.**

HOT TIP

Make sure everyone feels heard in your classroom! If 10 kids have their hands up to answer a question (wishful thinking, I know!), consider selecting three or four of them to share their thoughts instead of just one of them, and consciously make an effort to switch up who you call on! Thank the others for their willingness to share.

Proactive

It sucks that some of your students are walking into your classroom with potentially boatloads of math trauma that you had nothing to do with, but *you* can be the teacher who doesn't ADD to that pile. You can start fresh.

Starting tomorrow, you can begin to use Math Therapy to stop contributing to *more* traumatic math experiences and to avoid retraumatizing (or triggering!) your students. THAT. IS. A. HUGE. DEAL.

FYI, remember in Chapter 1 when I shared Gabor Maté's metaphor of trauma as a wound? You can think of a trauma trigger as something POKING that wound. The pokey thing itself doesn't have to be frightening or traumatic (e.g., maybe you just asked them to answer a question, or a test paper was placed on the desk in front of them) and may be only indirectly or superficially reminiscent of an earlier traumatic incident. Remember, when your student is being triggered, they are not necessarily responding to what just *happened* but to the feeling that lives in their body that is *triggered by what just happened*. Keep this in mind as you read through this chapter—it will help you understand what your students

> **When your student is being triggered, they are not necessarily responding to what just happened *but to the feeling that lives in their body that is triggered by what just happened.***

are going through and even what you yourself might be going through at times! And remember, the strategies that are used to help students heal from previous math trauma are the same as those used to prevent math trauma in the first place, which makes moderating trauma both a reactive AND proactive practice!

IN THE MOMENT

Have you ever had a memory triggered simply from a smell, sound, or photo? Reflect on your experience with triggers so that you can understand what your students are going through when they face math trauma triggers!

You know how I keep talking about how repetition and consistency are key to reinforcing concepts and beliefs? Well, that applies both to reinforcing new math concepts (yay!) . . . AND to reinforcing traumatic math experiences (boo!). Think about Paris again. She has a parent who has told her that boys are better at math, and that may have led to the formation of a math trauma for her. But now imagine that between Grades 1 and 9, she had a different experience. Maybe she was exposed to media that featured women being smart and mathy. Perhaps she found *Hidden Figures*. And the *Barbie* movie. And *The Mindy Project*! And *Lessons in Chemistry*. Paris's textbooks could have been full of diverse representations of mathematicians, including women she could relate to and identify with! Imagine Paris had a teacher in Grade 5 who made it super clear that there were *no* gender differences in math ability! In this scenario, by the time she got to your classroom in Grade 9 . . . she might be a VERY different version of Paris than I described previously.

We can't control what happens outside our classrooms or in our students' homes or what media our students consume or what math trauma they walk into our classrooms with. But we CAN control *our actions*. We CAN control what we choose to do tomorrow. And as the saying goes: Knowledge is power! The more we know about what our students are coming into our classroom with, the more we know what to do—and not to do—to avoid triggering or retraumatizing them, and the more we know what might help combat any negative messaging around math they may have received in the past. That is why I say that moderating math trauma is *so, so* important and is just as much about the past as it is about the future!

> **The more we know about what our students are coming into our classroom with, the more we know what to do—and not to do—to avoid triggering or retraumatizing them, and the more we know what might help combat negative messaging around math they may have received in the past.**

Before we get to our toolkit full of Actions and Activities, I want you to pause for a second to make sure that you're up to speed on both the potential causes of math trauma, as well as the way math trauma might present itself in your classroom. We went over all that in Chapter 1, but depending on how you're reading this, that *may* feel like a very long time ago, so do me a favor: Head back over to pages 12 and 21 and just do a little refresher to make sure we're all on the same page here! I'll wait.

BUT WHAT IF MY STUDENTS DON'T HAVE ANY MATH TRAUMA?

I recently did professional development (PD) with a group of middle and high school math and science teachers. Now, this PD was aimed at empowering educators to help their *students* heal their math trauma (just like this book!), not at encouraging educators to heal their *own* math trauma. That being said, I'm a strong believer in the idea that empathy for others can be strengthened by finding a point of connection. If we can truly understand where someone is coming from, if we can find a point of commonality, we are way more likely to empathize with them. So as I always do when engaging in this type of PD, I started out by asking if any of them thought they may have experienced math trauma *themselves*. Of the 30 teachers there, only three of them raised their hands. Most of them were confident with math and science, especially the high school educators—and while it's well documented that many elementary teachers identify

> **If we can truly understand where someone is coming from, if we can find a point of commonality, we are way more likely to empathize with them.**

as being anxious about math (Gilreath, 2023), the same can't necessarily be said for middle and high school educators.

As our PD progressed and we started talking about the multitude of events that might give way to a traumatic math experience, however, things started to shift. I explained that math trauma didn't have to come from huge, dramatic events—but that sneaky trauma occurs as a result of microaggressions, lack of representation, well-meaning comments about lack of ability, comparison to others, being yelled at for not understanding, timed tests, being shamed for getting the wrong answer or doing something the wrong way, and a host of other experiences that may seem relatively benign but that can lead to the development of a math trauma—especially when those experiences are repeated. I then asked them to find a partner and to consider again whether or not they may have had an experience with math trauma. When time was up and I asked who wanted to share, the tone had completely changed.

- A Grade 7 science teacher told us that she had always loved math but that in Grade 5 a teacher had given her failing marks on EVERYTHING because, even though she was getting correct answers, she wasn't doing anything "the teacher's way." She was so embarrassed that she didn't tell her parents and had to hold in the lie until her report card came out, at which point she broke down.

- A Grade 11 math teacher told us that he had never had an issue with math or science until he took this one particular college-level physics course. He had wanted to be an engineer. When he got to torque and velocity, he just couldn't wrap his head around the concepts. He ended up failing the course, and the experience totally scarred him. He became a math teacher, but to this day, he admits that he's TERRIFIED of science and refuses to even CONSIDER adding science to his list of teachables because of that ONE experience with those two physics concepts!

- A calculus teacher told us that while he had experienced many of the events I had listed (most notably, a parent who repeatedly yelled at him when he didn't understand a concept right away or get a perfect test score), that he had *not* experienced math trauma. In fact, he told us that these practices informed his teaching, that he now teaches math in the way that he experienced it, and that was that.

Almost every educator in that room had gone from "Nope, I've never had an experience with math trauma" to "OMG WELL LET ME TELL YOU ABOUT THIS ONE TIME . . . !!!!"

Not everyone who has experienced something that *might* cause math trauma necessarily has a traumatic experience of that event. Remember, math trauma is defined by the *experience* of the event, not the event in and of itself. At the same time, most of us don't realize that we've experienced an event as a microtrauma unless we have help moderating it! We're not here to insist that all our students have math trauma, but to create the space and strategies to explore their complicated feelings around math so that they can build a better relationship with it. Keep that in mind as you're working through this chapter!

Most of these educators *liked* math. They liked *teaching* math. But they—like all of us—were struggling with how to reach SO many of their students who seemed either disconnected, disengaged, disinterested, disheartened, or all of the above. And in this session, through our collective moderating of math trauma, a big part of the solution seemed undeniable: true, authentic *empathy*.

One misconception about math trauma is that it's reserved for those who lack math skills or who struggle with mathematical content. But that's not true! In her new book, *Math-ish*, Jo Boaler (2024) talks about how negative relationships with math are more common than we might think and don't only take place for those who don't identify as being "good at math." She says,

> I have been teaching highly accomplished undergraduates at Stanford for years now. Most of them arrive with a broken relationship with math. They have been successful, but they see math as a set of procedures they need to reproduce, at speed. When I show them that math can be the opposite of this—a set of connected and creative ideas that people can think about slowly—they are amazed, and thrilled. The students tell me that they do not ever want to go back to the narrow, speed-based version of mathematics they knew before. (p. 13)

Another misconception about math trauma is that once you've had an experience with it you are forever turned off math—you shut it out of your life, you can no longer even LOOK at it without having a total meltdown, and there is no f*cking WAY you would ever CONSIDER becoming a MATH TEACHER.

Sure, that's *one* possible end result of math trauma.

But just like math anxiety doesn't necessarily show up the way we expect it to, the same holds true for math trauma.

Our math traumas shape us as people—and as educators. And anxiety doesn't always look like panicking. And math trauma doesn't always look like math avoidance. In a second, we're going to take a look at some of the ways that math trauma might manifest in the ways your students show up in your math class so that you can get an idea of what to look for and how to spot it!

 IN THE MOMENT

I know I asked you this back in Chapter 1, but I'm wondering if your answer has changed now that you've read all this. Reflect on whether or not you think you have math trauma. What experiences do you think may have contributed to its development? How does it show up in your life right now? If you had math trauma and you feel like healing has taken place, what has helped you heal? Keep these experiences in mind as you head back into the classroom and consider sharing them with your students— as well as with your colleagues!

Now look, some of your students really might not even have a smidge of math trauma—it's totally possible! And as I mentioned earlier, the good news for YOU is that everything you're learning in this book is going to help you PREVENT math trauma from happening for those fortunate enough to have not encountered it yet. The point here isn't to convince every single person in the room that they *have* math trauma, but to validate those feelings should they exist and to create a culture in which math trauma is simultaneously healed—and prevented in the first place!

The following email just popped into my inbox from California math educator Kim Schultz, and I wanted to share it with you guys as a humblebrag, but also to illustrate how powerful it can be to just introduce students to the concept of math trauma and the stories that go along with it!

Hi, Vanessa!

I saw you present about a year or so ago and have been trying much of what you spoke about with my college students. This semester I pushed it further because I teach preservice teachers. I have focused on growth mindset with them for a long time, so it was a natural move. Here are the big things I have tried so far:

- *Teacher candidates started their semester with their math stories and will end the semester with their revised stories.*

- *I had my preservice teachers listen to your podcast episode, "What Even IS Math Trauma?" Their responses were as expected . . . triggered and at the same time validated. Most expressed their desire to change math education for their future students as well.*

- *With all of my students, I do mindset work and have many conversations about how "math people" don't exist; instead, everyone has the capability to do it.*

I have been trying to address traumatic practices in K–12 for years, and in the last 10 years, I have been teaching preservice teachers and trying to impact education that way. So I wanted to reach out and let you know you have hit on something so powerful! Thank you so very much, and I look forward to seeing how far you go.

BUT WHAT DOES MATH TRAUMA LOOK LIKE?

We don't have 10 hours for me to go through every potential cause or consequence of math trauma, but I want to remind you that math trauma shows up in our lives—and in our classrooms—in so many ways. In what follows, I have expanded on some of the info I gave you in Chapter 1 to remind you of some of the potential things to be aware of!

CAUSE	EXAMPLE	MIGHT SHOW UP AS	ANTIDOTE
Lack of representation	Textbooks that contain photos of only white male mathematicians	Statements that suggest a sense of not belonging—like, "I'm just not a math person," or "People like me can't do math."	Expose students to diverse resources and examples!
Nature vs. nurture	Being told that some people have the "math gift" and others don't	Lack of interest, expressing defeat when concepts take a while to understand	Emphasize that growth mindset is the real deal!
Public embarrassment	Being put on the spot to answer a question you don't know the answer to	Freezing when asked a question, "hiding" to avoid being called on, adversity to risk-taking	Continue to cultivate a Math Therapy classroom where kids work collaboratively and empathize with one another

(Continued)

(Continued)

CAUSE	EXAMPLE	MIGHT SHOW UP AS	ANTIDOTE
Pressure to perform	Being told that high grades in math are the only way to be successful	Freezing on an assessment, spiraling negative thoughts about grades, perfectionism	Help kids unpack the meaning of "success" so that it encompasses more than just grades
Labels	Being labeled "gifted" or "slow" or anything that suggests ability is fixed	Shutting down when faced with struggle, defiance, acting out	Avoid fixed language that implies that anyone (you included!) were just "born this way"
Lack of support	Feeling as though there is no way to get help when struggling	Opting out, absenteeism	Check in with students regularly to make sure they're feeling supported

Of course, not every single person has math trauma, and math trauma isn't the source of every single behavior your students exhibit. But I wonder if we can begin to view how our students show up *through* the lens of Math Therapy. Can we begin to perhaps see that behind chronic absenteeism, disinterest, underparticipation, anger, or what we often label "laziness" or "apathy" that there *might* be something more? That perhaps the root of these behaviors might actually be math trauma? If we can open our minds to seeing that, with empathy and compassion, we can begin to help our students *heal*. And that's what we're going to do next!

> *Not every single person has math trauma, and math trauma isn't the source of every single behavior your students exhibit. But I wonder if we can begin to view how our students show up through the lens of Math Therapy.*

HOT TIP

Start thinking about your class list—have any of these stories inspired you to rethink student behaviors from the perspective of math trauma? How can you bring empathy, compassion, and curiosity to those behaviors and the students who display them?

Look around! In addition to the many causes of math trauma listed in Chapter 1, there is often a great deal of math trauma caused by what is known as "the hidden curriculum." This refers to the implicit messages kids get from our education system. For example, kids can def be negatively impacted when they see gifted programs that have fewer Black kids than white kids (Sparks, 2022) or that the robotics team has only, like, one female-identifying student in it. Be on the lookout at your school for situations like this that might be impacting your students so that you can mitigate them with messaging in your classroom!

PUT *MODERATING* INTO PRACTICE

In the last chapter, we talked about how important the 3 B's are to building a better relationship with math:

Believe. Behave. Become.

These 3 B's are going to follow us throughout the 5 steps of Math Therapy, and I want you to always keep them in mind while working through the Actions and Activities to come. Why? Because the Actions and Activities I'm going to share will always be a mixed bag, intended to help your students not only change the way they *think (believe)* about their relationship with math, but to take steps to *change (behave)* their relationship with math so that they can feel better about math . . . and themselves *(become)*. Now, because this step is all about moderating math trauma, our toolkit is going to make use of some life-changing, transformative paradigms that I have been using to heal myself—as well as my students—for well over a decade! If you're, like, "I really don't care, just get to the Actions and Activities already," I totally understand. Who doesn't love a shortcut? I promise, this won't take long, and just like it's important to teach our students *why* a certain method works, it's just as important for me to teach you *why* the Actions and Activities in this toolkit are so effective!

Mindfulness and Math Trauma

I find that when most people hear the word "mindfulness," they immediately think of, like, hippies meditating in a commune, on a mountaintop far, far away. And, yes, while mindfulness does have ancient roots, it has gained widespread recognition and acceptance in the West through secular applications, therapeutic interventions, and scientific research demonstrating its positive effects on mental well-being. In

2017, I actually took a mindfulness-based stress reduction (MBSR; Teasdale et al., 2014) course that changed my life—and my teaching practice. I will be sharing some of my favorite MBSR strategies that have been adapted for Math Therapy in the Step 2 Toolkit, and yes, meditation IS one of them!

Now, while meditation *is* a mindfulness *technique*, it is definitely not the be-all and end-all of mindfulness! Mindfulness encompasses a vast array of strategies and tools designed to help us simply *be here now*. It sounds simple, but paying focused attention to the present moment is almost impossible for many of us (not just our students). Mindfulness is about being aware of your thoughts, feelings, and surroundings *without judging them*. It is this focused awareness that allows us not just to feel calm in moments of stress, but to really *witness* what it is we are feeling and experiencing. It is this combination of stress-reduction and focused, calm reflection that makes mindfulness such an important part of moderating math trauma, as it allows students to identify and process their emotions in a way that makes room for healing.

IN THE MOMENT

Do you have a mindfulness practice? Are you open to developing one? Joining your students on their journey to being more mindful is a great way to show them that you're willing to watch your brain grow and change right along with them and that you're all in this together! You might consider taking 5 minutes out of your morning or evening to set a timer and meditate in silence to get started. You can then share your experience with your students and maybe even inspire them to do the same!

Cognitive Behavioral Therapy and Math Trauma

I cannot overemphasize how much cognitive behavioral therapy (CBT) has helped me equip students to heal their math trauma. In fact, I published a series of kids' math books in 2017 called *Math Hacks*. Now, these books were mostly focused on helping kids work through the Canadian Grade 3–6 math curriculum, but I decided that I was going to write the first set of math books EVER that actually places an emphasis on the mental health of the students reading it. To do that, I decided to include an entire section of Math Therapy exercises that I created for kids, adapted from CBT techniques I had learned over the years. To this day, I still get emails from parents telling me that their child literally uses my MATH BOOK as a BEDTIME BOOK because they love reading the pep talks and doing the little exercises.

The thing is, CBT sounds so fancy, but really, it's just like having a personal detective for your thoughts and behaviors. While mindfulness is about being in the present, CBT takes it a step further by empowering you to pause, identify negative thought patterns or beliefs, challenge them, and ultimately replace these unhelpful thoughts with more balanced and constructive ones. It's not JUST about reflecting—it's about actively changing the way you think to improve how you feel and behave! CBT is such a great tool for moderating trauma because it really helps students dig into their thought patterns, where these patterns come from, and how to ultimately change them moving forward!

 IN THE MOMENT

If you have ever had a therapist, consider whether any of the techniques you learned in your sessions with them might give you insight into how to expand your Math Therapy skillset in the classroom! For example, CBT techniques I learned with my therapist have definitely made their way into my teaching practice, and I have even included some of them in the toolkit I made for you!

Now that you have an idea of two of the major paradigms that influence your toolkit, let's get to the actual Actions and Activities you're going to use to help students moderate their math trauma!

YOUR *MODERATE* TOOLKIT

Remember, the toolkit for each of the 5 steps of Math Therapy includes both *Actions* (which are quick) as well as *Activities* (which take a bigger time commitment). If you *can* find the time to squeeze in an Activity or two, it will definitely augment the Math Therapy process. But even if you only have the chance to try the Actions, they will have a huge impact on helping your students  heal their math trauma and build better relationships with math. As always, simply start where you are and do what you can—it WILL make a meaningful difference!

As you start to build a Math Therapy classroom, the *Moderate* Toolkit will become more and more effective. Remember to start by letting students know that *most* people have had negative math experiences that have shaped the way they feel about math. This is a message that should be repeated often so that it becomes normalized. And remind them that math trauma isn't something to be ashamed of: It is *normal*, just like all of the HUNDREDS of uncomfortable experiences that shape us as people.

Moderate Actions

Use the following Actions as a part of your regular classroom practice to help students moderate their math trauma!

 ### Moderate *Action 1: Create Mindful Moments*

Why

Guide your students to learn how to redirect themselves from a triggered state to a calm one through mindfulness!

Remember back in Chapter 1 when I told you that when students are anxious and steeped in their trauma responses that they literally CANNOT learn anything? Well, this Action is a quick way for you to help students get into a state that can support learning! By calming the mind and focusing on the present, we can make a huge difference in how our students show up in math class!

When

Start every single class with a mindful moment (it literally takes 1 MINUTE!) and incorporate mindful moments into your lesson whenever you feel your students need it. I'm a huge fan of throwing mindful minutes in both before AND after assessments—it makes a major difference and allows students to moderate their emotions quickly and efficiently, in the moment, every time.

How

1. When students are seated and ready for class to begin, set a timer for 60 seconds. Instruct students to plant their feet on the floor, sit up straight, and close their eyes. If they're uncomfortable closing their eyes, have them choose an object to focus their gaze on. All they have to do for 60 seconds is focus on their breath, in and out. If thoughts arise, suggest they observe those thoughts as though they are simply clouds passing in the sky. Don't judge them or engage with them, just watch. It may seem like a short amount of time, but trust me—60 seconds makes a WORLD of difference (Norris et al., 2018). Teachers have told me that the first week of doing this is tricky, but as mindful minutes become a consistent start to class, students begin LOVING them, even asking for more of them! If you have time, try using the same technique to end your class every day to de-escalate tricky emotions that may have popped up over the course of the lesson.

2. Sprinkle in mindful moments throughout your lesson! Mindful moments are ALWAYS a great way to bring kids back to the present from a place of potential trauma-response territory. Remember we talked about how moderating can mitigate the side effects of math trauma, which often include panicking, blanking on a test, or going into a negative thought spiral about math ability? When your students' emotions have taken them so far off the path that they seem unreachable, a mindful moment can literally help redirect them back to the present moment! Here are some of my favorite mindful moments that you can sprinkle in as needed:

a. **Mindful Observation:** If you ever ask your students, "What do you notice, what do you wonder," you've already got a head start on mindful observation! By asking students to pause, be present, and simply observe what a shape, photo, or illustration makes them notice and wonder, students are brought back to the present moment. Consider using everyday objects or interesting photos for mindful observation. Place an item in the center of the room and ask students to observe it closely, noting its colors, textures, and shapes. This exercise promotes focused attention and acts as a gateway to exploring mathematical concepts without the pressure of getting a right answer! Additionally, if a student is stuck on a problem and begins to panic, simply ask them to pause and ask, "What do you notice, what do you wonder!"

SOURCE: Canva

b. **Mindful Breath Counting:**
Focusing on our breath is a tried, tested, and true method to help us reduce anxiety by focusing on the present (Cho et al., 2016) because our breath is ALWAYS with us. Plus, usually when kids experience a trauma response like anxiety, their breath becomes irregular, causing their body to go into panic mode. By having students count their breath, their breathing slows down, helping them instantly feel more "normal." Scientifically, engaging in slow, deep breathing is a way to activate the parasympathetic nervous system, promoting a relaxation response. Here's a simple breathing exercise to stimulate the parasympathetic nervous system:

Box Breathing

Find a comfortable position

Have kids sit upright and place their hands on their lap or place one hand on their chest and the other on their abdomen.

Inhale deeply

Instruct students to inhale slowly and deeply through their nose and encourage them to allow their abdomen to expand as they fill their lungs with air. Aim for a slow count of four as they inhale.

Hold it

Have them hold their breath for a brief pause, counting to four.

Exhale slowly

Finally, have students exhale slowly and completely through their mouth or nose, letting the air out gradually while counting to four.

Repeat

Continue this deep-breathing pattern for several cycles!

This breathing exercise stimulates the vagus nerve, which plays a key role in activating the parasympathetic nervous system. *Bonus*: It involves counting—yay, math!

Grade-Level Modifications

- **Grades K–2:** Start every day with 60 seconds of silence and use a fun, visual timer like a giant hourglass filled with sand to set the tone.

- **Grades 3–5:** Hear me out: glitter jars! Mindfulness and math expert Deborah Peart has this amazing activity that involves making meditative glitter jars that kind of look like snow globes, but with glitter! Kids then have their own mindfulness tool to use in times when a quiet moment is needed—all they have to do is turn their magical jar upside down and watch the glitter swirl and settle.

- **Grades 6–8:** Ask students to share their favorite techniques for relaxing in times of stress and consider sprinkling them in whenever you include a mindful moment during class!

- **Grades 9–12:** Encourage students to build a mindfulness practice outside of the classroom. Apps like Headspace offer free subscriptions to educators and go a long way in terms of helping kids build up their meditation muscles. Headspace (Headspace, 2017) also has mini pep talks and mindfulness lessons specific to school stress and exams!

⧗ Moderate *Action 2: Moderate Your Emotions!*

Why

Help your students moderate their math trauma by empowering them to identify and label their emotions! Kids aren't necessarily scared of math—they're scared of the *emotional consequences* of math! This Activity helps remove some of the fear around emotions by normalizing them and breaking them down into manageable, identifiable components!

Students often have complicated feelings around math that they have never learned how to identify. By helping students better understand their emotions, we can help them move away from the panicky feeling that often accompanies math trauma and move instead toward labeling emotions as they pop up, recognizing emotions are normal, and finding ways to manage specific emotions related to math trauma!

When

Include opportunities for reflecting on emotions as part of every test or assignment! Additionally, you might consider giving students 5 to 10 minutes at the end of each week to reflect on what emotions popped up for them in math class that week! (See the free template on the Math Therapy website!)

SENSE OF
FEELING

Name: _____

Circle an emotion you felt this week while learning or doing math

Confused **Disappointed** **Sad** **Embarrassed** Other: _____

Mad **Happy** **Stressed** **Proud**

What thoughts do I have when I feel this emotion?

How do I act when I feel this emotion?

THOUGHTS **ACTIONS**

FEELINGS

Where in my body do I feel this emotion?

Strategies I can use to manage stressful emotions if they arise

How

1. Talk about it! Most students are never given a chance to talk about their emotions. This is your opportunity to help them moderate their feelings about math so that they don't just feel encased by a giant cloud of frustration any time math comes up. Use the downloadable template on the Math Therapy website to crowdsource ideas from your students about what different emotions they feel when math comes up. Use the legend provided to have them describe what each emoji and emotion means to them. You can start by talking about each emotion as a class, as well as some coping skills that might work to manage complex emotions. Follow up by providing them with template Math Therapy journal pages to write their own descriptions since emotions will show up differently for everyone. Here is an example of what might come up in a group setting:

Sad **Sadness**

What thoughts do I have when I feel this emotion:

- Like I can't do it
- I feel hopeless and stupid
- Like maybe it would be better if I never did math again so I don't have to feel sad

Where in my body do I feel this emotion:

- My stomach
- Near my heart
- I feel like I can't breathe in my lungs

How do I act when I feel this emotion:

- Sometimes I shut down and try to be invisible
- I cry
- My mind goes blank, and I can't learn anything

Coping skills I can use to manage stressful emotions:

- Taking deep breaths
- Writing down how I'm feeling to let it all out
- Talk to a friend, parent, or teacher

2. Pop it in! Allow students to moderate their emotions at the end of any assessment, formative or summative. Provide them with an emoji legend, ask them to circle the emotions they experienced during the assessment, and considering asking the following:

    a. Circle the emotions you felt during this assessment.

    b. If you circled more than one emoji, pick one to talk about right now! Where did you feel this emoji in your body?

    c. What thoughts went through your head while feeling this emotion?

    d. Did you use any skills to manage this emotion? List them below!

Make sure you grade these questions as a part of the assessment so that students can see that they are just as valued as "correct answers."

3. Dedicate time! Set aside 10 minutes at the end of each week for moderating emotions. Students can use their Math Therapy journal pages for this exercise. For best results, collect these reflections and look them over, providing students with direct feedback and considering an adjustment to pedagogy as needed!

Grade-Level Modifications

- **Grades K–2:** Provide each student with a set of emoji cutouts that they can hold up or place at the front of their desk when they're experiencing a difficult emotion.

- **Grades 3–5:** Consider using a more limited set of emotions to start with so that students don't get overwhelmed by all the different choices.

- **Grades 6–8:** When doing your end-of-week reflection for moderating emotions, allow students to use illustrations as well as words to describe what emotions they experienced around math that week.

- **Grades 9–12:** When doing the initial crowdsource, let students use their phones to generate a word cloud that will literally illustrate the emotions being expressed! There are a ton of free word cloud generators—I love this one: https://bit.ly/3uQrJeX.

 Moderate *Action 3: Watch Your Language!*

Why

Choose your language carefully to avoid triggering and retraumatizing your students!

One of the ways in which math trauma can be triggered is through our use of language. Students who have experienced math trauma have likely heard certain trigger words before (e.g., "bad" at math, "easy" for everyone but you), and every time they get the impression that they're receiving a similar message, retraumatization can occur. Of course, all of us are well-meaning and are never intentionally trying to trigger our students, but we don't know what we don't know. That's why a simple and effective way to reduce the amount of triggering that happens in your classroom is to be aware of what words might trigger your students so that you can avoid using them and replace them with something better instead.

When

Literally all the time! In with the old, out with the new—swap out triggering words for nontriggering words any time you can!

How

1. Familiarize yourself with statements and words that may trigger your students and swap in something more asset-based and growth mindset-oriented instead. A key part of moderating trauma is figuring out how to avoid triggering and retraumatizing, and language is a crucial part of that. Here is a table to help you get started!

INSTEAD OF THIS	CONSIDER THIS	WHY?
"This is easy—anyone can do it."	"Everyone can improve their math skills with practice and support."	We don't want to give students the message that if something isn't "easy" to *them* that they're doing it wrong. Ease isn't the goal in math!
"You should know this by now."	"It's okay if you're still getting the hang of it; we all learn at our own pace."	Students feel bad when they feel behind, so showing them that speed and ability aren't correlated reduces stress!
"This is a hard concept—most of my students don't get it."	"If you ever feel challenged in my class, that means you're learning—enjoy the ride!"	The word "hard" is scary for many students! Replacing it with "challenge" is more of a growth mindset vibe and invites students into the process instead of shutting them out.

(Continued)

(Continued)

INSTEAD OF THIS	CONSIDER THIS	WHY?
"You should be able to solve this quickly."	"We value thoughtful problem-solving, not just quick answers."	Speed is one of the number-one causes of math trauma. As soon as we tell students they need to be fast, they shut down—try this instead!
"If you don't get it now, you'll never catch up."	"It's okay if you're finding this challenging; we'll work on it until you feel confident."	Kids have been hearing about learning loss for way too long and being labeled as "behind" or a "slacker" is traumatizing. Reinforce the idea that learning math isn't a race!
"I was always bad at math too."	"I find math challenging as well, but with practice, I'm starting to notice small moments of clarity!"	It's powerful to be vulnerable with your students and let them know you struggled with math, and it's even more powerful to make it clear that improving your relationship with math is all about practice and perseverance, not about innate skill!

2. Add to your list! If you notice other statements or keywords are triggering your students, add them to your list of things to look out for and prepare alternative statements or keywords that you can swap in. We all have different habits and verbal ticks (like, I say "like" ALL the time, in case you didn't notice!) and bringing awareness to those and having alternatives on deck and ready can help US help our students feel better in the moment!

HOT TIP

Don't be hard on yourself if you slip up and say something you didn't mean to! These are great opportunities to pause, reflect, and pick a different word next time. If you notice it in the moment, you can even say something like, "Guys, I just said this problem would be EASY, but I'm going to retract that and here's why!"

Grade-Level Modifications

- **Grades K–2:** Be specific and direct when making language choices and stick to words that don't have multiple meanings.

- **Grades 3–5:** Use simple, positive language without nuance so that there is less risk of misinterpretation. For example, instead of saying, "What an interesting approach," say something like, "What an interesting approach . . . I really like how you used creativity to solve the problem!"

- **Grades 6–8:** Consider creating an anchor chart where you, along with the class, define the most common terms you use. For example, when you say, "Do your best," what does that mean? Get everyone on the same page!

- **Grades 9–12:** Engage in a discussion about why language is important and encourage students to monitor their own language as well—both when talking to one another and to themselves!

Moderate Activities

Use the following Activities to engage students in moderating their math trauma on a deeper level!

 Moderate *Activity 1: Dear Math . . .*

Why

Engage your students in moderating their math trauma by having them talk about math in a familiar, personal, and nonthreatening way!

We're all familiar with writing emails, text messages, notes, and diary entries. This Activity is a creative, fun way to open your students up to talking about math in a nontraditional way that not only allows them to communicate how they feel but enables you to understand where their math trauma might stem from and how it shows up for them in your classroom!

When

This Activity is a one-time or two-time thing at most! Do it once at the beginning of the year and a second time at the end of the year to observe how your students' relationship with math has progressed!

How

1. Set the tone! Set aside time for this Activity by setting a timer for 15 to 20 minutes in class. Consider playing instrumental music, dimming the lights, having students clear their desks, and getting everyone in the zone to reflect and write.

2. Share an example! Explain to students that they will be writing a letter to math. There is no right or wrong way to do this Activity, and the only goal is to imagine that you're *actually* writing a letter to math: What do you want to say? Share some examples, which you can find on the Math Therapy website or in Sarah Strong and Gigi Butterfield's book *Dear Math*.

> Dear Math,
>
> I really want to like you...but you don't come naturally to me! I feel like I have to work harder than anyone else to actually GET what it is you're all about. There have been times when I've felt defeated, frustrated, and bummed out - especially on tests which is where I feel I get extra stuck. Even when I KNOW the answer, sometimes it just doesn't come out on paper in a way that makes sense.
>
> XOXO

3. Commiserate and investigate! Invite students to share their letters with the class. Collect the letters so that you can get an understanding of what each of your students might be experiencing when it comes to their complex relationships with math. These letters provide valuable

clues as to what math traumas your students have experienced and will help further inform your classroom practice!

4. Revisit! Consider revisiting this Activity again at the end of semester or year so that students have the chance to reflect on their relationship with math after going through the 5 steps of Math Therapy! It can be so interesting and rewarding for you to see how far they've come, AND as an added bonus, it can be even more rewarding for THEM to see how far they've come! Once they've submitted the second round of letters, have them examine their first and second letters side by side and note the differences. These letters provide tangible evidence of their growth over time, which is THE WHOLE POINT OF MATH THERAPY!

Grade-Level Modifications

- **Grades K–2:** Remember *Mad Libs*?! That's how you're going to do this activity with the littles! Create a letter outline with a ton of fill-in-the-blanks and put a bank of possible word choices at the bottom of the page so that they have easy access to options if they get stumped!

- **Grades 3–5:** This group is often the most uninhibited and . . . creative! Make sure you set a few parameters by letting them know that their letter should start with the words "Dear math . . ." and include a minimum of three sentences. This is usually enough to get the ball rollin', and you can encourage them to use illustrations in addition to words if that helps!

- **Grades 6–8:** Give students diverse options in terms of what shape they want their letter to take. Illustrations can help this grade band explore how they feel, so encourage them to use emojis or to create comic strips where the protagonist is having a conversation with a character that represents math!

- **Grades 9–12:** Allow students some autonomy regarding the modality of their final product. Instead of a letter, students might choose to make a confessional video or a blog listicle titled "Ten Things I Love (and Hate) About Math!"

 Moderate *Activity 2: The "Ick"*

Why

This Activity is all about introducing students to the concept of math trauma in a light, easy, and relatable way so that they can begin to feel validated and understood. You don't even need to use the words "math trauma" if you don't want to. This Activity will help students give their feelings a name so that they can identify them when they pop up. (Check out the free downloadable on the Math Therapy website!)

HOT TIP

Know your audience! If "trauma" is a word that's already being thrown around your classroom by students, you're fine to straight-up bring up the concept of *math trauma*. If that isn't a word that your students use, refer to it as "negative experience with math" or "uncomfortable experience with math" or "the ick." Play to your crowd, you know them best!

When

This is a great way to introduce the idea of math trauma to your classroom, and I recommend doing this in entirety and then making space to revisit Steps 2 through 4 once a month. You'll find that as you start to build a Math Therapy classroom that students are more likely to be able to spot and identify uncomfortable feelings and experiences around math, label them, and talk about them. When you repeat Steps 2 through 4, you can reframe them to be more current (i.e., In the past month, have you had any uncomfortable experiences with math?).

How

1. Name it! Start by introducing students to the concept of math trauma. If you're working with younger students (Grades 3–8), you might consider using "icky math feelings" instead of "math trauma." All that is required in this initial step is that you introduce students to the idea that many of us have icky feelings around math and that we're not alone. It's perfectly normal to feel as if you have a complicated relationship with math! Remind them about what you talked about while you were mythbusting and reiterate the message that we're not born with icky math feelings, but over time, many of us develop them. List some examples of what might lead to those feelings! You can bring up examples from your own life, refer to the stories included in this chapter, or head back to Chapter 1 for a refresher on what might cause math trauma!

2. Time to *Moderate*! If you feel like your classroom vibe currently allows for it, have students get into pairs and make a list of ANY situation or event they can think of that made them feel icky about math. If you think this won't go well in your classroom, no worries—have them open their Math Therapy journals and make their list in there! If they're working in pairs, have them chit chat to come up with their list, and then get them to each write their list contents down in their Math Therapy journal. They might write about the time their parents

compared them to a "smarter" sibling or the time they did a timed math test and totally froze up.

3. Explore it! Now that their lists are made, they can go a step further in the moderation process and write out how each of those events made them feel. Did they feel bad about their math abilities? Did they feel unsure of themselves? Whatever they remember feeling—write it down! They might write that when their parents compared them to a sibling, it made them feel as if they weren't good enough. And when they froze on a timed test, it made them feel like they weren't smart. If your class is working in pairs, have them discuss and then write the results down in their journals.

4. Talk it out! If you can encourage students to share their icky math experiences and the icky math feelings that resulted, that is the perfect way to end this Activity. Sharing the ick is super cathartic. It not only helps students feel heard, but it almost always leads to validation, as you'll hear a lot of "OMG, the same thing happened to ME" from the rest of your students. As students share, make sure to emphasize that these math traumas that they experienced aren't their fault and that they don't define them as students or people. They are just one part of their relationship with math, but just like any delicate relationship with a friend, there are ways to make it stronger and better, and you will be helping them work on that!

5. Leave it open! Reiterate that you know the whole idea of math trauma (or "icky feelings") might be new to them and that you're proud of them for digging deep and looking at their math history from a new perspective! Let them know that you are committed to helping them rebuild their relationship with math by providing better math experiences and that in a month you'll be checking in to see how you did! When the time comes, circle back and repeat Steps 2 through 4!

HOT TIP

Do a temp check! It's helpful for students to know they're not alone and that math-related events that have caused them discomfort have caused others to feel the same way. Just make sure you're not trying to convince anyone that every math experience they've had *has* been traumatic! You can do this by asking instead of telling. So for example, saying something like, "When I was younger, I had a really bad experience with timed tests, and ever since then, I freeze up in math class. Has anyone else had that experience?" is a great way to introduce an event example while opening it up to individual students to decide how they feel!

- **Grades K–2:** At this stage, focus on making sure students don't feel isolated if they experience ick around math. You might do this by talking about your own math story or emphasizing that many people have complicated feelings around math and that doesn't need to define them as a math learner!

- **Grades 3–5:** Leave the term *math trauma* out of it and get THEM to make up a name or sound to describe uncomfortable math experiences! What kind of sound or word might be used? They might say something like "SHPLAT" or "PFFFT" or who knows? Kids are random! Let this be a fun part of the process that allows you to dig deeper into what negative math experiences feel . . . and sound like!

- **Grades 6–8:** It's likely too soon to use the term *math trauma* for these grades, too, but you CAN talk about math anxiety and stress when you're exploring the ideas of "icky" feelings or discomfort!

- **Grades 9–12:** This grade band lives on social media and is very used to hearing the words "trauma" and "trigger." If you have the bandwidth for it, feel free to go a little more in depth about what a math trauma or trigger might look like and how it might affect the way they show up in math class and how they might feel about themselves as math learners in general!

Moderate *Activity 3: The Math Therapy Classroom Agreement*

OUR MATH CLASS CONTRACT

We promise to be guided by these values:

RESPECT:
- We will listen and follow directions.
- We will take turns.
- We will be open to everyone's ideas.
- We will address each other politely.

COLLABORATION:
- We will actively participate in group activities, and work together as a team.
- We will help others when they are struggling

KINDNESS
- We will encourage each other.
- We will offer a helping hand when needed.
- We will celebrate mistakes
- We will make sure everyone feels included
- We will be patient if one of us needs the teacher to slow down
- We will apologize if we hurt someone's feelings

We pledge to do our very best to follow these agreements, inside and outside of the classroom.

Elias   Alexandria   Lucy   Connor   Michael   Liam

Why

Create space for students to moderate their trauma by further cultivating a Math Therapy classroom with a classroom agreement that enables everyone to feel respected! (There's a free download on the Math Therapy website that will help you with this!)

 **IN THE MOMENT**

Make promises you can keep! We talk a lot about creating "safe spaces" for students, which is a beautiful goal—but one we should be careful with. A safe space is a place or environment in which a student can feel confident that they will not be exposed to discrimination, criticism, harassment, or any other emotional or physical harm. While we WANT to provide safe spaces for our students, the question is this: CAN we? Stay away from calling your classroom a safe space and instead refer to it as a Math Therapy classroom. You can stipulate what that means—for example, discrimination will not be tolerated. That's as close as we can get to guaranteeing students that they won't *be subject to* discrimination because that's simply not a guarantee we're equipped to make! A safe space is the goal, but until we get there, strive for a Math Therapy classroom!

There is nothing I love more than a math agreement, and I love the emails I get from educators showcasing the variety of ways this can take shape. By co-creating a "contract" with your students, you show them that student voice is valued, that respect is a two-way street, and that you are committed to making sure you all work together to create more positive math experiences! An added bonus is that you can gain an understanding of what math traumas are present in your classroom based on what students say!

When

Do this once as close to the beginning of the year as possible. Refer to the Math Therapy classroom agreement as needed and consider revisiting it once a month to edit and revise as required!

How

1. Discuss! You might talk about what contracts are and how we use them in the real world! The point of a contract is to outline rules that are beneficial to both parties—in this case, your students and you! Let them know that the goal here is to create more positive math experiences for EVERYONE, and you need them to work together (and with you!) in order to do that.

2. Crowdsource! It's time to figure out what to put in your agreement. Now remember, a lot of this will be about compromise from BOTH you and your students. Here are some questions you might ask to start gathering information for your agreement:

a. What do you like about math in a group?

b. What do you dislike about math in a group?

c. What is your favorite part about math class?

d. What is your least favorite part about math class?

e. If you could change one thing about math class, what would it be?

Actively listen to what your students say and make notes on the board so that they can see that you are taking their thoughts seriously!

3. Build the agreement! People will have to compromise because not everyone will agree. For example, someone might say that what they dislike about math in a group is getting picked on when they haven't raised their hand. Other students will likely agree. You might concede to not pick on anyone without a raised hand *in exchange for* a promise that each student will find a way to participate once a week. BAM! Now you can put that on the agreement! The goal is to come up with a Math Therapy classroom agreement that allows everyone to relax a little bit, feel that they are being seen and heard, and open up to building a better relationship with math thanks to tangible evidence that you, the teacher, are making compromises to help them on their journey!

4. Frame it! Okay, you don't have to frame it literally—but you should put the final agreement up where everyone can see it. These are like the commandments of your classroom! If they aren't being followed (by you OR your students), you can respectfully point to the "commandment" that has been violated. This is a great way to talk about how we can communicate our feelings and express our emotions in kind, respectful ways, and how we can respond to one another in situations where conflict may arise. Overall, these agreements go a long way in cultivating a Math Therapy classroom and can be referred to over and over again to set the vibe as needed—consider having students take turns reading the agreement out loud to start your lesson every day (at least until everyone is used to the rules)!!

 **Grade-Level Modifications**

- **Grades K–2:** Provide examples of what might appear in a classroom agreement to get students going.

- **Grades 3–5:** Consider turning your math agreement into an art project everyone can contribute to! Write the agreement on a giant poster board and then have each student contribute a cut out illustration that relates to Math Therapy! Hearts, doodles, brains, and stick figures are ALL welcome!

- **Grades 6–8:** If students are feeling shy, invite them to contribute their answers to your prompts on anonymous slips of paper. Collect them all and then read them out loud to the group. You'll find that multiple students will likely write similar statements, which will help direct your discussion!

- **Grades 9–12:** High school students can be savvy negotiators—watch out! A Nevada administrator told me that when she does this with her classes, she adds an additional prompt: "What do you think I, as your teacher, would like to see in this agreement?" That opens up the door to talking about mutual respect and having students be more willing to compromise WITH YOU, as well as vice versa!

MODERATE TOOLKIT: SIMPLE SWAPS

Want to take moderating to the next level? The six classroom practices in the table that follows have been shown to cause math trauma in many of our students (Boaler, 2015; Owen, 2021). Don't worry if you use some of these—we all do! The key thing is to know that they might lead to trauma and to adjust accordingly *where you can*. Sometimes we *have* to give students a timed test or teach them an algorithm—we don't necessarily have full autonomy over how we assess or the curriculum we teach. As with everything else in this book, start where you are and do what you can! To help you out, I've included the following simple swaps to effectively sub out some of your current practices for those that will help you *moderate* your way to a Math Therapy classroom!

SWAP THIS	FOR THIS
One-shot assessments	Give students multiple chances to take the same assessment at different points during the year so they can be accurately assessed for the knowledge they have accumulated and mastered.
Timed tests	If students need extra time, let them have it.
Group students by ability	Put students in randomized groups of three or four (Liljedahl, 2021).
Only valuing one way of solving a problem	Allow students to use different methods to get to the same solution.
Focusing solely on memorization	Emphasize understanding and critical thinking. Encourage students to explore the "why" behind mathematical concepts.
Strict "right" or "wrong" grading	Give marks for process in addition to getting the right answer.

MODERATE TOOLKIT: THEY SAY, YOU SAY!

It's always hard to come up with responses on the fly, so here are some of the things I suggest in response to some of the most common statements we hear from our students every day when it comes to moderating math trauma!

WHEN THEY SAY . . .	YOU SAY . . .
"I don't feel like doing this."	"I totally get it. What can I do to help you do this even though you don't FEEL like it?"
"I'm stressed."	"Stress is a totally normal response to something that feels hard or to something that you care about! Stress and anxiety have nothing to do with ability, so let's figure out how to work through those hard feelings together!"
"I feel stupid."	"It sucks to feel that way, and it's totally not your fault!" (This is where I personally would go on a ginormous rant about how our culture makes us feel like the only way to be smart in math is to never make mistakes and how this is total BS . . . !)
"I suck at math."	"You don't suck at math! You've just been made to feel like you do, and I'm sorry. Tell me more about why you think you suck at math, though—I'm curious!"

 ## *MODERATE* TOOLKIT: EXPANSION PACK

I have so many more *Moderate* Activities that I have shared with teachers over the years, but there are only so many pages in this book . . . off to the website you go! The following are a list of more of my fave Activities. You can find templates and instructions for most of these Activities by heading to the Math Therapy website!

1. **Make Representation Happen:** Many of your students will have experienced math trauma related to a lack of representation in all of the places in their lives where math happens. That might be on TV, in textbooks, and even in the classroom. You can help moderate their math trauma by making a point to include diverse mathematicians in your activities and conversations, using resources from diverse educators, and talking about the many ways in which folks can be good at math! For more on how to incorporate culturally responsive

pedagogy in your classroom, check out *Choosing to See* (Seda & Brown, 2021)—it's one of the best books out there!

2. **An Attitude of Gratitude:** Developing a gratitude practice has been cited as one of the best ways to rewire neural pathways so that our thoughts moderate to a more optimistic and positive place (McDermott & Spann, 2022). Usually, when we come from a place of math trauma, it's hard for us to not automatically fixate on the worst-case scenario! Check out the downloadable template for some extra Math Therapy journal pages you can use to help your students cultivate feelings of gratitude throughout the week. Some educators start each of their classes by asking students to list three things that they're grateful for that day. This really puts students in a positive place to begin learning! (Remember what I said earlier? Students have to be in the right STATE to learn, so bringing them to a place of thankfulness and calm will help them—and you—in the long run.)

3. **The "What If?" Game:** One of my favorite activities to lead students through is what I call the "What If?" game. Have you ever heard the saying "Prepare for the worst, hope for the best"? Well, so have our students—and it sucks. Seriously, why are we teaching kids to spend their whole lives preparing for the worst-case scenario?! That is NOT the vibe we want to cultivate in a Math Therapy classroom. Why? Because our students are already wrapped up in imagining that the worst is yet to come—they'll make a mistake, fail a test, end up jobless and single forever . . . no. They do NOT need any more practice at that. What they need to practice is imagining "what if?!" WHAT IF the BEST-case scenario happens? This activity is simple: When a student expresses their fear over something going horribly wrong in math class (failing a test, getting a bad mark, getting the wrong answer), ask them instead what the BEST thing is that could happen in that situation. Have them describe it to you, visualize it, close their eyes, and imagine how it FEELS—and THEN do the task they've been avoiding due to fear. It will make a difference! Shifting the energy in any given moment opens students up to de-escalating a potential trauma response and taking a chance!

4. **Play the Script Until the End:** This technique is especially useful for those struggling from fear and anxiety related to math trauma. If a student is feeling this way, engage them in a thought experiment in a one-on-one setting. Ask them to imagine the final FINAL outcome of the worst-case scenario (e.g., I'm going to put my hand up to ask a question, and the whole class will laugh at me, and then . . . ?). Letting this scenario play out can help your student to recognize that EVEN IF everything they fear comes to pass, the outcome will still be manageable—they will move on just like they have in countless other similar scenarios!

5. **The Body Scan:** One of the most effective mindfulness exercises for helping students get back to the present moment is the famous body scan. This is GREAT for a student who is having a *moment* in class or in the middle of a test and needs help chilling out. Body scans are simple!

- Have students sit upright, plant their feet on the floor, hands in their lap.

- Ask them to focus ONLY on how each part of their body feels, one part at a time. The key is to start at the feet and slowly work up to the head. Suggest spending 10 seconds on their feet and toes, moving up to the calves, kneecaps, and so on. As your students begin to focus their attention on their physical body, their thoughts meander from their negatively spiraling thoughts—to their tangible, physical presence. By the time they reach their head, they are usually fully present, their breathing has regulated, and they are able to get back to the task at hand.

STEP 2 TOOLKIT
TO GO
moderate

moderate ACTIONS

1. CREATE MINDFUL MOMENTS

Try box breathing, 60 seconds of mindfulness to start class, and "what do you notice, what do you wonder?"

2. MODERATE YOUR EMOTIONS

Help students identify their emotions, track how they feel, and understand how emotions affect behavior.

3. WATCH YOUR LANGUAGE

Choose your language carefully to avoid words or phrases that might be triggering.

moderate ACTIVITIES

1. DEAR MATH...

Have students write letters to math to moderate their math trauma!

2. THE "ICK"

Help students give their feelings a name so they can identify them when they pop up.

3. THE MATH THERAPY CLASSROOM AGREEMENT

Cultivate a Math Therapy Classroom by creating a classroom agreement!

moderate EXPANSION PACK

MAKE REPRESENTATION HAPPEN ✳ AN ATTITUDE OF GRATITUDE ✳ THE 'WHAT IF' GAME ✳ PLAY THE SCRIPT UNTIL THE END ✳ THE BODY SCAN

BEFORE WE MOVE ON

Another step of Math Therapy under your belt, which means it's time to celebrate with a cute selfie, moments of reflection, and of course—a treat!

Treat Yourself

Time to treat yourself to a little mindful R-and-R to celebrate your progress! Take a moment to do something that helps you feel fully present. Put on your fave song and dance like no one's watching, eat a delicious snack without any distractions so that you can focus on how great it tastes, or watch a juicy movie without checking your phone even once! Whatever you do, be here now and feel the relief of having everything else fade away into the background!

ASK YOURSELF

1. How did this chapter help you moderate some of your own math trauma that may have been lurking below the surface?

2. Which interpretation of the word "moderate" resonates with you most?

3. Are the ideas presented in this chapter different from your current educational philosophy? If so, how?

4. Which of your students do you think will benefit most from the Actions and Activities suggested in this chapter? Which do you think won't benefit as much? Why?

5. Which Action or Activity are you most excited to try with your students?

6. What is something in this chapter that makes you go, "aHA, this is totally going to make a difference"? Why?

7. What is something in this chapter that makes you go, "That will NEVER work with my students!" Why?

8. What are some of the challenges you may face when trying to implement the suggestions in this chapter? How might you prepare for these challenges ahead of time?

☮♡π

CHAPTER 5

MATH THERAPY STEP 3

Motivate Your Students

THE 5 M'S OF
MATH THERAPY

STEP 1: Mythbust

STEP 2: Moderate

✓ STEP 3: Motivate

STEP 4: Makeover

STEP 5: Measure

In this chapter, we get to:

1. Examine why it's important for students to connect to the skills they're being asked to build

2. Discover that there's no such thing as an unmotivated student

3. Learn how motivation actually works and how to harness it

4. Explore how pride and purpose play key roles in student motivation

5. Dig into a toolkit full of Actions and Activities designed to help motivate your students to want to build a better relationship with math

When Annie showed up at my tutoring studio, her parents told me that she was failing math, didn't care, and at this rate would have to take Grade 10 over again. I decided to spend the first session figuring out what Annie was all about, so instead of focusing on math, we focused on why she didn't care about math.

It turns out that Annie didn't care about math because she wanted to be a model. So like, why did she need the quadratic equation? This seemed like a fair point to me, tbh, as did her other points, which were that (a) she found math boring and (b) she feels like sh*t every time she does it, so why bother? It was pretty hard to argue with these solid points, so I didn't bother trying. Instead, I asked her about what skills were needed to become a model. We discussed that useful skills might include possessing confidence, being able to monetize modeling so that she could actually make a living, knowing her worth so that people couldn't take advantage of her, having stamina for long hours for photo shoots, and developing a dedicated skincare routine.

Since repeating Grade 10 math would have delayed graduation and, as a result, possibly delayed her potential modeling career, I suggested we focus on just passing math class instead of trying for high marks, which took a lot of the pressure off. I also suggested we double the payoff and work on mastering the skills she would need for modeling WHILE working on math. Surely, getting better at math would help her eventually monetize her career; planning her homework schedule would help build the same skills needed for organizing and sticking to a skincare routine; and getting more confident at anything—including math—would for sure help boost her sense of self in all areas of life! She was like, "Okay, you're weird," but said she'd keep it all in mind.

Once we started working on the actual math, I was SHOCKED. Annie picked up on new concepts so quickly and not just because she was a good memorizer, but because she actually understood the way I was explaining things. As I started to express my amazement at how mathematically savvy she was, she was like, "You're literally the first person who's ever said this to me."

Most of our sessions were sort of a combo of us arguing over how she hated math and then her acing the math I taught her. Her first test since she started with me was taking place after our third session together, and we decided she would focus on feeling good about how far she'd come and just try to pass.

The next afternoon, Annie came literally MARCHING into our studio after she wrote her test. I hadn't expected to see her, but THERE SHE WAS, ranting at the top of her lungs about how THIS question almost stumped her but she figured it out, and THAT question reminded her of something we had done together but there was a TWIST to it but SHE hadn't been fooled. I had NEVER seen her so excited! I told her I truly didn't care what mark she got on her test because THIS WAS THE WIN: her, in my studio, knowing her sh*t and caring about it was THE SUCCESS.

The next day she got her test back and got an 85%. This is the girl who literally had a 33% in her class. I called both her parents to scream my excitement, and they were genuinely confused. This was not the Annie they recognized! And they were *right, it wasn't. This was the Annie who was finally told that someone cared about what SHE thought and what SHE wanted to do with her life. This was the Annie whose voice was respected and valued and listened to. This was the Annie who, years later, is a model and no longer claims she's "not a math person." This is the Annie who commented on one of my TikToks yesterday and messaged me after to say that all of my social media posts about math trauma have helped her realize that she for sure had math trauma from years of being told she sucked at math but that she's living her best life and is confident in her math skills (but also still thinks I'm weird, even to this day).*

<   **All activity** ⌄

 **Annie Carter**
commented: Hearing u say the
quadratic formula gave me ptsd
😭 😭 bahahahahahahahahah 1m

View reply   ♡ Like

Do you have an Annie in your life? We probably all do, and although they challenge us, they implore us to dig below the surface-level responses we give students when asked WHY they should bother building a better relationship with math. By valuing student voice and helping students connect math to what truly matters to them, we motivate them to WANT to keep going on their Math Therapy journey—even when it may seem easier to quit. THAT is what Step 3 is all about!

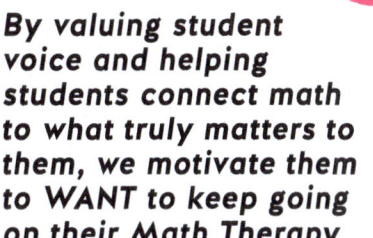

By valuing student voice and helping students connect math to what truly matters to them, we motivate them to WANT to keep going on their Math Therapy journey—even when it may seem easier to quit.

STEP 3: *MOTIVATE!*

Imagine this: Your students now know and believe that their brains are capable of doing (even enjoying!) math. They now understand that everyone has complicated feelings around math, that they're not alone in feeling this way, and that those experiences and emotions don't define them or what they're capable of, and they also have strategies to manage those complex emotions. But now you've got to convince them to CARE (if they don't already!). *Spoiler*: Not every student in your classroom is going to care about building a better relationship with math regardless of if they believe they can or not. The *Motivate* Toolkit you'll dive into later

in this chapter is all about giving students the motivation they need so that when they start doubting themselves (which we ALL do sometimes!), they can be reminded that they have a bigger reason for sticking it out!

 IN THE MOMENT

Think about something you know you can do but aren't necessarily motivated to do. What is that thing, and what would need to happen for your motivation to change?

 MOTIVATE

The third step of Math Therapy is all about how to *truly* motivate your students to *care* about building a better relationship with math. Change is hard and uncomfortable for literally everyone, whether it has the potential to lead to a better outcome or not. If we want our students to commit to putting in the effort, we need to convince them the payoff will be worth it. In this step, you will make use of impactful classroom strategies to raise the stakes for your students in a way that feels meaningful and valuable to them!

Why It Works: It is crucial that at the midpoint of our 5 steps, we get *buy-in from our students* if we don't have it already. We can have the best content and pedagogy in the world, but if our students aren't motivated to learn it, it ain't going to happen. When we talk about motivation in a Math Therapy context, we are taking a departure from traditional motivational tactics that focus on extrinsic rewards (grades, prizes, validation from adults in the room) and focus instead on intrinsic validation (pride, purpose, meaning, relevance), which we are rarely taught to tap into in the classroom. This step works because the *Motivate* Toolkit transforms valuable research about how true motivation *actually* works into practical strategies you can easily integrate as part of your current classroom practice.

I've been talking a lot about the importance of cultivating a Math Therapy classroom, and I want to emphasize that all the hard work you've been doing is going to continue to pay off in this step, big time. As your students start to feel as though their growth and progress matters and that their feelings and relationships are being nurtured, they will be more likely to embrace this step. Remember, it's all about taking a student-centered approach to finding out how to motivate your students, based entirely on *what matters to them*. By cultivating a Math Therapy classroom, you're already halfway there (sort of—I mean, Step 3 is mathematically the midpoint, so . . .)!

How It Works: The Step 3 *Motivate* Toolkit is full of strategies you will use to discover how to raise the stakes in math class so that your students truly feel motivated to build a better relationship with math! It will also equip your students with Actions and Activities that get them closer to feeling meaning, purpose, and value in your classroom, which in turn will empower them to take on challenges and take ownership over success. In tandem, your understanding of what actually motivates your students, coupled with their increased motivation to build a better relationship with math, make Step 3 powerful!

Like all 5 steps of Math Therapy, *Motivate* isn't a one-and-done technique, but instead it is a set of ideas and strategies that will help you *build* and *maintain* a Math Therapy classroom. Some of the tools in your toolkit will be things that you can do *once* with your students, and others will need to be used *consistently* and *repeatedly* with your students.

Before we get to our toolkit, we REALLY need to talk about how motivation actually works and how we can harness its power in our classrooms. I'm not going to be all, "I have found THE SECRET to motivating EVERY STUDENT," but I am going to be bold and say that I think I have discovered some *pretty* groundbreaking stuff that's going to potentially transform the way you motivate your students from this moment forward. A bold statement, I know, but hey—keep on reading and see if you think if I'm right!

 IN THE MOMENT

Up until now, what are some of the techniques that you have used to motivate students? Have they worked? Why or why not?

THE SECRET TO TRULY MOTIVATING YOUR STUDENTS

Okay, I promised earlier that I wouldn't get all "clickbaity" and claim I had solved some giant mystery, but here I am doing just that (couldn't help myself). Motivating our students truly does remain one of the biggest math class mysteries out there. They could legit make an entire true crime detective podcast series about this, and I feel like we would ALL tune in, but since that doesn't exist yet, you're stuck with me, and I'm here to tell you that the secret to motivating your students has to do with two things: understanding *how* motivation actually works and understanding *what* motivates our students.

What If My Students Are Unmotivated?

Over the years, I have had many educators say something along the lines of "I REALLY want to try Math Therapy in my classroom, but it probably won't work because my students are just *so unmotivated*." I feel you. Sometimes it seems as if there is *no point* in making the effort to try anything new because your students seem unreachable and just don't act like they care. But what if I told you that there is NO SUCH THING as an unmotivated student? Hear me out for a sec.

I am currently very unmotivated to go to the gym. Quite frankly, I would be the first one to call myself an unmotivated "athlete." But the thing is, most of my life I have legit been a gym rat. I have had a workout routine since the age of 12, and whether it was hitting the gym, going for a run, crushing a spin class, or doing yoga, I have always worked out pretty consistently. Want to know why?

Vanity.

You heard me. I grew up in the age of the thigh gap (really hoping that trend is so long gone that you're reading this now being like, "wtf is a thigh gap?!" Don't Google it, trust me), so my motivation came from the fact that I was always trying to lose weight, no matter WHAT I actually weighed. It was definitely body 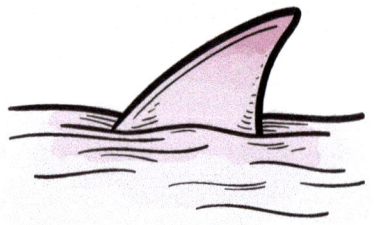 dysmorphia, but for the sake of this conversation, I call it "motivation." Now, it wasn't the GOOD kind of motivation—it was totally unhealthy and f*cked up—but it got my butt out of bed and to the gym, I can tell you that much. And this is going to sound controversial, but this whole thing isn't *too* dissimilar to how we used to "motivate" our students to learn math not so long ago. Think about it:

> **"Motivating" women to work out in the 1990s:** telling women that heroin-chic (Rosser, 2010) is the new societal standard to aspire to and that thinness is correlated with success.

> vs.

> **"Motivating" students to learn math in the 1990s:** telling kids that being good at math is essential for survival, that good grades are correlated with financial success, and that if they don't do well in math class that their privileges will be taken away and that they'll be basically grounded for life.

Why am I on this tangent? Well, for two reasons. The first is that I want to point out that the word "motivated" isn't inherently *good*. It's only as good as the intentions behind it, so when we talk about motivating our students, we're not talking about getting them to care by threatening or bullying or scaring them. We're talking about convincing them to care by tapping into *what's important to them*. These are two totally different things, so I just want us to all keep that in mind. I think we can all agree that if we try to use fear tactics to motivate our students, we will likely end up in the same mess that we're in now: a situation in which our students are *more anxious around math than ever before* (Dowker et al., 2016; Flannery, n.d.). Motivating our students with fear leads to—you guessed it—MATH TRAUMA! So no, that's not what we're doing here. The second reason I'm on this tangent is because I started this whole thing off by suggesting that there was *no such thing as an unmotivated student*. And I'm about to prove my point.

So I'm currently very unmotivated to go to the gym (as I just told you). I also told you that, 5 years ago, I finally got sober. Well, when I got sober, something totally unexpected happened: My body dysmorphia disappeared. At the time, I thought I had just made the decision to stop drinking, BUT it turned out that getting sober was ACTUALLY all about releasing the hold that addictive patterns have always had in my life, and apparently, many of the habits I had formed over my lifetime were simply a manifestation of my addictive tendencies. I was an addict—who knew! It turned out that I wasn't just addicted to alcohol, but I was addicted to working out and losing weight. At the SAME time that all this was happening, Kim Kardashian became super famous, and beauty standards went from "thigh gap" to "all the curves." I know this seems weird, but the combination of the two factors rendered me totally . . . unmotivated . . . *to work out*. The factors that had motivated me *had changed*, rendering me unmotivated, *but only temporarily*.

FIGURE 5.1   MY BAND PLAYING TALL PINES MUSIC FESTIVAL

A year later, my band was getting ready to go on our first tour. We would be playing 25 shows in 31 days and driving across Canada. Sounds totally glam, right? But what this really means is that we would be getting little to no sleep, spending 10 hours in a van every day, and playing an insanely high-energy rock show almost every night because we're just that kind of band—we scream, sing, jump, dance, and *don't stop* for 60 minutes, and we leave it ALL on stage. Playing one show is tiring enough, but there was NO WAY I was going to be able to play 25 shows in ONE MONTH if I wasn't in WAY better shape. So guess who was suddenly motivated again? THIS GURL.

See what happened there? I went from motivated, to unmotivated, BACK to motivated, and now I'm back to *un*motivated.

SOURCE: Allan Fournier

I am not an *unmotivated student*. I am *a student* who, at times, is unmotivated.

Do you see how powerful the difference is in that statement? It's the difference between the idea that motivation is either fixed (the former statement) or flexible (the latter statement). It's the difference between believing our students *can change* and believing they *can't*. It's kind of a big deal.

♥ IN THE MOMENT

Until I learned all of this, I totally thought that some of my students (and friends!) were simply unmotivated people. Go through your student list and ask yourself if there are any students you might have similar thoughts about. Are you reconsidering? Why or why not?

Here's the deal: Motivation is not static. It can change based on a whole range of factors, and we HAVE to remember that a student who appears unmotivated in one context might demonstrate motivation in another. We HAVE to hold this in our minds at all times because we have learned to dismiss certain students as innately unmotivated, and it's just not a thing. That thinking ends now!

> *Motivation is not static. It can change based on a whole range of factors, and we HAVE to remember that a student who appears unmotivated in one context might demonstrate motivation in another.*

How Does Motivation Actually Work?

New research on motivation shows that one of the key factors to whether or not a student is motivated lies in making sure they're in the right headspace—or in science-speak, the right *state*. There are certain states that are conducive to motivation and those that are not. In fact, boredom, fear, and anxiety are all states that are ABSOLUTELY NOT conducive to motivation (McConchie, 2022), which is why my eyes roll into the back of my head every time someone suggests that we should go back to teaching math the way we did in the 1990s.

Some of the states that ARE conducive to motivation include curiosity, confidence, hopefulness, optimism, empowerment, self-efficacy, satisfaction, mindfulness, gratitude, resilience, and open-mindedness. Liesl McConchie explains, "When a learner is in one of these biological states, the brain releases a cocktail of neurochemicals (such as dopamine, norepinephrine, and cortisol) that activate the motivation and learning systems of the brain (Corbett et al., 2017). These factors take place regardless of age, culture, or subject matter" (McConchie, 2022, para. 17). As you can see, we've already tapped into some of these states in the previous chapters so you're already likely ahead of the game in terms of getting your students into the various states that lend themselves to motivation. In this step, we're going to learn how to effectively activate even more of these states to authentically *Motivate* our students!

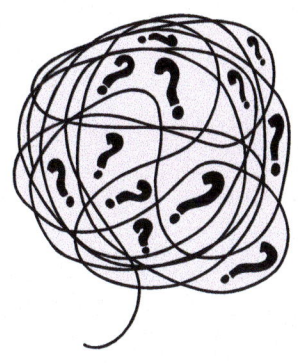

HOT TIP

The following hacks can help you quickly get your students into two of the *Motivated* states!

- **Anticipation**: To get your students into an anticipatory state, try covering a manipulative with a sheet for the first half of your lesson so students are left wondering what's underneath it.

- **Enjoyment**: Have you heard of something called a flow state? It's essentially when someone is "in the zone," fully immersed in and focused on an activity, and time just passes by—we've all been there! To get students into a flow state, the key is to give them math tasks in which the challenge is *just right* for their skill level. Since this can be tricky to do in a classroom with students of varying skill level, try using what Alicia Burdess calls "Big Beautiful Problems!" These are problems that are low-floor, high-ceiling, and accessible to everyone. In Alicia's words, "Big, Beautiful Problems are the ones that have the potential to change lives. They are the ones that can make you fall in love with math and see it differently than you've ever seen it before." You can check them out at www.aliciaburdess.com.

There are countless ways to get students into states that lend themselves to motivation, but I have consistently found that, by focusing on *purpose* and *pride*, I am usually able to cover multiple motivation-friendly states, ultimately motivating most students to care, *even if just a teeny, little bit* about improving their relationship with math. *Purpose* is about how to motivate students through helping them find meaning in what they're doing, and *pride* is all about how to motivate students through empowering them to take ownership over success.

Your Step 3 Toolkit will contain a ton of Actions and Activities centered around those two themes, so let's talk about them for a sec!

IN THE MOMENT

Watch out for demotivators! Just as there are ways to help our students get motivated, there are certain factors that are chronic demotivators (McConchie, 2022). These include systemic inequities such as school policies that inadvertently lead to a bias in tracking racialized students more than others or that involve districts receiving differing amounts of funding based on their socioeconomic demographics, leading to more supports for students in certain schools and less in others. Also included are factors unique to each student, such as fatigue from poor sleep, sickness, or acute stress—and of course, math trauma. Hold this in mind as you continue to cultivate a Math Therapy classroom and remember that you can't control everything, so focus on what you CAN control and on understanding the many factors that affect each of your students when it comes to finding motivation to build a better relationship with math!

HOT TIP

Use math mystery to get your students motivated to find out why the math works! Here is one of my faves:

The 1089 Trick

1. Choose a three-digit number where the digits are not all the same and arrange the digits to make the largest and smallest possible numbers.

    Example: Choose 352. Arrange to get 532 (largest) and 235 (smallest).

2. Subtract the smaller number from the larger one.

    Example: 532 – 235 = 297

3. Reverse the result and add it to the original result.

    Example: 297 + 792 = 1089

No matter what three-digit number you start with (as long as the digits are not all the same), if you follow these steps, you'll always end up with the number 1089. It's a fun and surprising trick that involves a bit of mathematical magic!

"But When Will We Use This irl?"

Ask any math teacher what question they get asked the most and it'll likely be, "But when will I use this irl?" or some derivative of that sentiment.

I believe this is one of the most misunderstood questions in math class—and one that has led to far too many textbook problems about curved bridges, spaceship flight paths, and the plight of some poor person carting, like, 456 watermelons home from the grocery store.

The thing is—our students aren't asking us when the particular math concept they're learning can potentially be used by a human on this planet. What they're really asking is when will THEY actually use this concept in THEIR ACTUAL REAL LIFE, like, NOW.

HOT TIP

Collaborate with school counselors or coaches to provide guidance on how to help your students find purpose! They can offer resources and tools for self-discovery and goal-setting and guide you to how you and your students might access those within your school!

"But when will I use this irl?" = "But why should I actually give a f*ck?"

Distilled into one word, what they're asking about is *purpose.*

Students need to feel like the skills they are learning in your class serve a *purpose.* Immediately! For skills to serve a purpose to your students, those skills must be relevant to them either now, or in the very near future! Students don't benefit from forced attempts to apply math to like, planet earth, *in general.* Take questions such as "There are 2 trains moving toward each other at 60 km an hour; they're 5 miles apart. How long will it take for them to pass each other?" or "Use sin law to figure out what angle a cable needs to be attached to the top of a roof in order to tether it to the ground." They might be "practical applications" of math, but guess what? NO ONE CARES, thus rendering such questions as ones without purpose!

Purpose in your classroom can be found in two central ways to help motivate your students: *math* skills and *math-adjacent* skills.

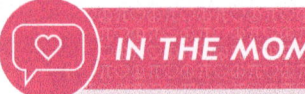
During one of the reflection periods you offer your students, ask them to journal about what matters to them. What do they enjoy, what do they find meaningful, and what gives them a sense of purpose? Have them submit these reflections for you to look over and see if you can find a way to incorporate what you find into your classroom practice!

Meaningful Math-Adjacent Skills

Math-adjacent skills are skills that are built by doing math but that aren't necessarily considered exclusively mathematical in nature—many of which, like **"reason abstractly and quantitatively"** and **"make sense of problems and persevere in solving them"** are built into the math standards. For example, when working on a challenging problem, students might try multiple methods to solve that problem, work with friends to get new perspectives, or make a mistake and start again. They are building math-adjacent skills like risk-taking, critical thinking, collaboration, and learning how to grow from mistakes. These are all *meaningful skills* that can be used *in their real lives, right now!* They can use these SAME skills when making new friends (risk-taking, collaboration), trying out for the soccer team (risk-taking, collaboration, growing from mistakes), or getting a part-time job (risk-taking, critical thinking, collaboration). Emphasizing that the math-adjacent skills they learn in your classroom will enhance their actual lives RIGHT NOW is a simple and effective way to motivate your students.

HOT TIP

The next time students ask you when they will "use this irl," highlight the math-adjacent skills they are learning instead of just focusing on the math!

vaneSSa vakharia
@TheMathGuru

Math teachers: how do you answer the question "but when will I use this in real life?"

Nicole Ronald (she/her) @MissR... · 21h
I tell them it's not specifically the math content they will use in real life but the skills like problem solving, communication, and perseverance that they will absolutely use in real life!

Linda Bortnick @lbmb62997 · 14h
I tell them math reaches parts of the brain other subjects don't, so when they're learning a skill what they're actually doing is developing their thinking skills and you need to think "in real life!"

Tim O'Connor @timmyo40 · 1d
Love this question: I'm training your brain to be able to do mental gymnastics for whatever unusual/unexpected/crazy thing you see next. I'm helping you think creatively, consistently and critically. Will you ever factor polynomials? Maybe not.

Melissa D @Dean_of_math · 1d
I show a clip of Neil DeGrasse explaining that it's about the process of learning the math, not the math itself.

Meaningful Math Skills

It is A LOT harder to convince your students that the math skills they are building in your class serve a purpose (you probably know this from experience!); therefore, you have to work A LOT harder to use this as a motivation tactic! That doesn't mean it can't be done, but I totally get how hard it is, given that we have little to no control over the actual math content we teach.

What this means is that we need to find creative ways to connect this content to our students' actual, lived experiences. One way to do that is by treating our content in culturally responsive ways. This might involve recognizing and incorporating diverse cultural perspectives, experiences, and examples into your curriculum so that students see themselves reflected and find the content meaningful *to them* (Matthews et al., 2022). Another way to show students that the math skills they are learning serve a purpose in their lives is by using relevant examples (remember, relevance means it *matters to them right now*). Esther Brunat (n.d.) is a math teacher who is a total pro at doing this! Over the years, I've seen her students learn patterning and probability by exploring the latest trends and fads (like, why are so many people suddenly wearing Crocs again? Or how likely is it that in a group of 100 that at least one person will be wearing this hideous footwear, and does this stat differ by region?), learn all about money by creating their own imaginary small businesses, and learn how to graph by filming TikTok dances to identify how different equations visually move (and yes, she used this activity as an approach to *summative* assessment, which is genius if you ask me!).

Finally, one of the most effective ways to bring purpose to the actual math skills students are building in your classroom is by broadening the definition of "math" to include all of the skills we tend not to emphasize enough—skills that actually matter to your students. These include skills like estimating, trendspotting, navigating, budgeting, managing time, logical reasoning, making change, optimizing, making comparisons, and literally solving ANY type of problem. When we talk about math in this way, kids have the opportunity to connect the math they're learning to what brings them meaning, and also, kids are usually able to at least find ONE skill that they feel capable of and confident about, which is a HUGE win when it comes to *Motivation*!

So last summer, my band and I went on tour across Canada for 5 weeks. Now, let me back up and tell you that my band is made up of me . . . and four dudes—one of whom is an engineer and three others who profess to HATE math and suck at it. Eric, our guitarist, is one of those three. But here's the thing: the five of us drove across the entire country, crammed into a SUBURBAN, and every night, after every show, had to load ALL of our gear (instruments, amps, suitcases, merch, everything) into that Suburban's trunk, a challenge most would find basically impossible (me included, tbh).

But night after night, there was Eric, packing that car trunk like he was playing 3D Tetris, irl. I mean not an INCH to spare, everything squeezed in PERFECTLY to maximize the space of that trunk. Literally people would stop on the street and stare in disbelief, placing bets on whether or not he would be able to do it, and night after night, he did. He was literally maximizing three-dimensional space in a near-impossible circumstance, ALL THE WHILE professing that he sucked at math, and I kept being like, "ERIC, YOU ARE DOING MATH. THIS IS MATH. I DO NOT CURRENTLY POSESS THIS MATH SKILL AND WOULD NEVER KNOW WHERE TO EVEN BEGIN WITH FITTING ALL OF OUR SH*T INTO THIS TRUNK."

And, guys, the craziest part was the LOOK he would get in his eyes when it was time to pack: This crazed, excited GLEAM would take over, and he would get SO excited for the challenge that he would throw on these construction gloves, disappear into some sort of geometric fugue state, and just GIV'ER. By contrast, packing a trunk is literally my worst nightmare (that's just not the kind of math I'm into, tbh). Like at one point, in the midst of his precision packing, he popped his head out of the trunk to survey all of the gear lying around on the sidewalk and literally yelled—I'm not kidding—"Hey, someone hand me something trapezoid-shaped!" Like he said the word "TRAPEZOID." IN REAL LIFE. If that's not math, I don't know what is (see Figure 5.2).

At some point during our tour, I made a TikTok video of Eric doing crazy car-trunk-packing-math (https://bit.ly/3Tj4nIg), and The Field's Institute (an international center for scientific research in mathematical sciences in Toronto) saw it and reposted it with the caption, "There are many ways to do math!" Eric literally called his mom to tell her. It was the first time he had ever been told by a "school" that he was anything more than worthless in math class. Even all these years later, that mattered. That *meant* something.

FIGURE 5.2

SOURCE: Eric Robbs

HOT TIP

To incorporate cultural responsiveness into your Math Therapy classroom, try these things:

1. Encourage collaborative and cooperative learning strategies.

2. Use resources and teaching materials created by a diverse range of educators.

3. Provide opportunities for students to share their perspectives in math discussions.

4. Use varied assessment methods.

5. Be aware of and address potential cultural biases in instructional materials.

What Is Success Anyway?

I have major beef with the dictionary.

I have tweeted *Oxford Languages*, like, 72 times, petitioning them to CHANGE their definitions of "success" and "failure," to no avail. That's fine—I have a book now, so I get to be in charge of the discourse around success and failure (at least within the realm of these pages anyway).

According to our pals at the *Oxford Languages* (Oxford Dictionary, 2023), success and failure are essentially defined as antonyms:

SUCCESS	FAILURE
1. The accomplishment of an aim or purpose	1. Lack of success
2. The attainment of fame, wealth, or social status	

We'll get to that second definition of success in a second, but let's just focus on the *first* side-by-side definitions of failure and success. HOW WILD IS THAT??? How sad—and inaccurate—is it that we are taught to define failure and success in opposition to one another when the reality is that the two are an essential *part* of one another.

Success is defined as the accomplishment of a "goal," but when I ask students how many times they've set out to accomplish a goal, have *failed*, yet something WAY better happened, almost ALL of their hands go up! The last time I did this with kids, one brought up that she hadn't gotten into her first-choice college, but on her search for a backup plan, she found a college program she was WAY more excited about! Another student said he hadn't made the soccer team, so instead, he started taking karate lessons and was now working toward his black belt! I mean, if Netflix's *Love Is Blind* has taught us literally anything, it's that failure is sometimes THE BEST thing that can happen to us! Hello, how many couples go on that show to attain the goal of getting MARRIED (their version of success), yet end up getting ditched at the altar (their version of failure), then get interviewed on the reunion show, being like, "OMG, getting ditched at the altar was THE BEST THING THAT EVER HAPPENED TO ME! I dodged a MAJOR RED FLAG that could have RUINED MY LIFE." Hello! Does that sound like the OPPOSITE OF SUCCESS TO YOU? I'm just not having it, and quite frankly, I know you aren't either; but the thing is, this is the discourse that students are dealing with in basically EVERY area of their lives, so it's up to us to change it! My two fave ways to do JUST that are coming up next, and they involve both *Mythbusting* success and inviting students to take ownership over what success means to them!

 **IN THE MOMENT**

Pause and consider how you define "success." What does it mean to you? Now, ask yourself how you define success for your students. What is the same about the way you defined both of those? What is different?

Invite Students to *Mythbust* Success

Oh snaaaap—I just did a throwback to Step 1 because that's the whole point of Math Therapy! All these steps work together and have a cumulative effect on busting math trauma.

Mythbust success by making it clear that failure and success are like the key ingredients to everyone's fave salad dressing, oil and vinegar. Seriously, every science teacher wants to tell you that those two things don't mix well, but the truth is that your salad is going to taste pretty gross if you choose to only include one *or* the other! While they might separate at times, they're better together baby!

HOT TIP

Remind and re-remind your students that failure and mistake-making are not the opposite of success but are an essential part of the learning process. Give them examples of famous mathematicians who have taken years to solve problems and scientists who have taken just as long to find cures to diseases. Why? Because they were spending all that time FAILING AND MAKING MISTAKES!

Invite Students to Take Ownership Over Success

One of the most powerful ways we can help our students to take pride in their work is by allowing them to define *what it is that makes them proud in the first place*. Students have been taught that the only way to be "successful" in math class is by being externally validated by us, the teacher. But they're getting mixed messages! Don't we ALSO teach students not to care what others think? To do what feels authentic to them? That self-love and self-efficacy are more important than a mark on a piece of paper? That INTERNAL validation is way more potent and powerful and personal than EXTERNAL validation?! We need to merge these two disparate discourses around success, and I think most of us can agree that the latter is what we want our students to focus on.

Deepak Chopra explains that "it's common to sometimes measure our success or fulfillment by comparing ourselves to others. This is called object-referral, looking outside for validation. It's also common to over-look the opportunities for success that exist in our setbacks or failures . . . However, self-referral, looking within, is where true fulfillment resides" (The Chopra Center, n.d.).

SELF-REFERRAL. INTERNAL VALIDATION. *THIS IS WHERE TRUE FULFILLMENT RESIDES, GUYS.*

> **One of the most powerful ways we can help our students to take pride in their work is by allowing them to define what it is that makes them proud in the first place.**

I know I'm yelling, but it is IMPERATIVE that we give students the opportunity and means to *redefine success so that it is meaningful for them.* I'm not sure how many of you had parents with a certain idea of what success should look like for you; many of my students do. I'm thinking of Jessie (not their real name), whose parents keep insisting that being a dentist or doctor are the only two ways to be successful in this life. As a result, they obviously put a lot of pressure on her to do well in math class and are sorely disappointed to discover that she not only isn't doing well, but doesn't seem remotely motivated to do any better. Why? Because Jessie wants to be a creative director at an ad agency, not a dentist or doctor.

This is a classic story, one I'm sure many of you are familiar with, but it points to the crux of the problem: HER PARENTS have defined success in a way that is NOT meaningful to Jessie; therefore, she is not in any way motivated to achieve that success. When I asked Jessie what success might actually look like to her, she told me that she didn't care about grades necessarily. She hated the feeling of blanking out on tests or sitting through a 60-minute lesson feeling confused. So I invited HER to create her OWN definition of success. The definition she came up with was this:

Success to me means that I understand what my teacher is saying and that I feel smart.

Notice that everything she wrote down was about HER, not about how others—her parents, teachers, friends—PERCEIVED her. Once she had a definition of success that meant something for her, she was motivated to strive for it, and she felt an authentic sense of pride in doing so! What's more exciting is that after a month she no longer felt confused in class and even felt "kind of smart," and she asked me if she was allowed to edit her definition of success to build on how far she had come. As you can imagine, I was all, "HECK YES, THAT IS THE POINT!" Success is not stagnant—it changes as we grow and learn! This ALL goes back to Step 1, guys! Remember "I used to think, now I think?!" Jessie decided that her NEW definition of success was going to be this:

"Success to me means that I understand what my teacher is saying, I feel smart, and I try at least 50% of the questions assigned to me."

And so it went for the remainder of the year. Jessie now works at one of Toronto's coolest ad agencies, but guess what she does on the side for fun? SHE TUTORS MATH. Not even kidding. It's basically a rom-com ending!

I want you to remember why you picked up this book. It wasn't just to help your students build a better relationship with math so that they could get better grades or pursue a STEM career . . . it was to help your students build a better relationship with math *so that they could lead happy, healthy, amazing lives full of wonder and possibility*! What could be more in line with that goal than teaching a young person how to define success for themselves in a *meaningful* way?! In a world increasingly obsessed with defining success in terms of the assets we accumulate or status we achieve or likes we get on our latest Instagram post (see that second definition of success I threw up there in the chart) . . . this is the stuff that really matters, guys!

At the end of the day, when we give our students the autonomy to redefine success, they are more likely to care about achieving "it," whatever that is. And that's what *Motivation* is ultimately all about: caring.

HOT TIP

Ask your students what success means to them and engage in a discussion about how success is allowed to be different for everyone!

PUT *MOTIVATION* INTO PRACTICE

This step is an exciting opportunity to get to know your students even more than you already do! You've been tapping into what makes them feel uncomfortable around math, which often means dredging up icky emotions, but in this step, you get to tap into what brings them hope, joy, inspiration, excitement, and *MOTIVATION*! This is a FEEL-GOOD toolkit, full of Actions and Activities to get your students amped up. As you continue to build a Math Therapy classroom, you are going to feel the vibes start to rise in your classroom—it's a feeling like no other, and I can't wait for you to tell me ALL about it!

YOUR *MOTIVATION* TOOLKIT

Ready to raise the stakes in your classroom in a way that feels meaningful to all students? Let's do this! As with every Math Therapy toolkit, I'm giving you both *Actions* (which are brief and don't take much prep) as well as *Activities* (which take more time). As always, simply start where you are and do what you can!

Now, let's get MOTIVATED!

Motivate Actions

Use the following Actions as a part of your regular classroom practice to get students motivated to build a better relationship with math!

Motivate *Action 1: Assign Accountability Buddies*

Why

When students feel accountable to some-one else, it gives them an added sense of purpose!

Have you heard of the Blue Zones? They're the five places in the world where studies have consistently shown that people live the *longest, happiest lives*. Of course, researchers have put a lot of work into studying WHAT makes these places so special, and it boils down to 10 specific habits and practices, including *purpose and meaning*, which is defined as engaging in meaningful activities, having clear goals, and a sense of belonging and connection in your community (Blue Zones, n.d.). Feeling accountable to another person and like you *matter* in your math classroom is a great way to motivate students with *purpose* while also adding to the culture of your Math Therapy classroom!

In case you're wondering, the five Blue Zones are Okinawa (Japan), Sardinia (Italy), Nicoya (Costa Rica), Icaria (Greece), and Loma Linda (California), and I plan to travel to them ALL to see WHAT IS UP!

♡ IN THE MOMENT

A sense of belonging can play a crucial role in whether students are moti-vated in your classroom or not. The concept of "belonging uncertainty" (Walton & Cohen, 2007) describes the uncertainty students might feel about their belonging when entering new academic situations, most notable during times of transition (e.g., entering high school or college). Research has shown that belonging uncertainty affects how students make sense of daily adversities, often interpreting negative events as evidence for why they do not belong. If a student feels like they don't belong in the world of math, for example, then they are more likely to interpret mistake-making or being told they're wrong, as an affirmation that they aren't a "math person." You can address belonging uncertainty by focusing on classroom practices geared toward helping all students feel like . . . they belong! For example, Georgia educator Kirk Lunde places emphasis on student autonomy and respect, meeting students' sensory needs, showing his students diverse examples of mathematicians, and making it a point to positively affirm every student equally. Now that you know that belonging uncertainty is a thing, think about how you can help all students feel like they matter, they're worthy, and . . . they belong!

When

Assigning accountability buddies only needs to be done once! The key is for students to have the same accountability buddy throughout the entire semester so that they can build a collaborative and trusting relationship and really learn to count on one another. Throughout the year, I suggest dedicating 10 minutes a week to having students do a check-in with their accountability buddy during class time.

How

1. Explain! Begin with a brief discussion on the importance of collaboration and support in learning. Acknowledge that most of our math experience has been all about riding solo and that actual mathematicians not only work on solving the world's biggest problems together, but they have one another to pep talk in times when they feel like they're just not getting anywhere or to help strategize with when they don't know the best approach to something. Ask your students what the benefits might be of having a go-to person they can count on in math class.

2. Set expectations! Clearly communicate the expectations for accountability buddies. This includes the frequency of check-ins, the nature of collaborative work, and the overall purpose of the partnership. A good place to start is by explaining that accountability buddies are the ride-or-die in your classroom. They're the person who might do the following:

   a. Take notes for you when you miss class

   b. Remind you of an important deadline

   c. Ask you for help if they forget how to do something

   d. Give or receive a pep talk when it's needed most

The whole point of your accountability buddy is to support you and to get support *from you!* I saw Sarah Strong, author of *Dear Math* (Strong & Butterfield, 2022), do a presentation where she talked about how one of her students FaceTimed her accountability buddy into a math class lesson when she was home sick, so that she could learn along with everyone else! Ask kids to contribute to the list of what accountability buddies might do for each other and let them know that, once a week, they will be given 10 minutes to meet with their accountability buddy for a check-in!

3. Assign accountability buddies! I've heard from teachers who have done this in both randomized and nonrandomized ways. I tend to think that randomized works better simply for the fact that kids know that their pairing was random and don't jump to erroneous conclusions about why you paired them with a certain person (Liljedahl, 2021).

4. Meet and greet! Give students 10 minutes to meet up with their assigned buddy and to talk about what they would like to get out of the arrangement. What do they need support with? How do they want to communicate (phone, email, Snapchat, whatever app is trending at the time of you reading this book . . .)? Is there anything they need to know about one another's learning style or study habits? Some students want to text and FaceTime and do homework together; others are fine with just checking in once a week. Let students have flexibility over their arrangement and check in with them individually after 2 weeks to see how it's going.

5. Weekly check-ins! Give students 10 minutes at the beginning of every week to sit down with their accountability buddy and use the following prompts:

   • How did you feel about what we learned last week?

   • Was there anything you found confusing or stressful?

   • Was there anything you felt really good at?

   • Is there any specific way you want me to support you this week?

These prompts give each buddy the chance to jump in with a pep talk or to discover that their partner struggled with something that *they* actually felt successful at, so they have the chance to help! This sense of purpose will motivate them to continue to show up on their Math Therapy journey even if it gets hard.

6. Involve the class! After the first 2 to 4 weeks, consider having pairs share the successes of their buddy system with the class so that other pairs can get ideas about how to further strengthen their accountability bond. Encourage them to share challenges they may have faced and how they overcame them, as this reinforces growth mindset and further cultivates the trust and collaboration you're building in your Math Therapy classroom!

Grade-Level Modifications

- **Grades K–2:** Keep accountability buddy duties limited to in the classroom by asking students to focus on classroom compliments and check-ins!

- **Grades 3–5:** Tell students exactly what they're supposed to do with their accountability bud. In this grade band, it's most effective to ask kids to give their accountability buddy compliments (make sure you share examples of what an effective compliment looks like!), help their accountability buddy get help if they don't understand a math concept, and remind their buddy about upcoming deadlines. Kids don't all have phones so make sure you set up time in class for accountability buddies to check in with one another!

- **Grades 6–8:** Students in this grade band may or may not have consistent access to a phone, so set up accountability buddy check-ins at the beginning of each lesson. All you need is 5 minutes for students to check in with each other using the prompts in #5 in order to build morale and get a sense of where everyone is at in the room!

- **Grades 9–12:** High school students can be shy or self-conscious when it comes to asking for help from a peer. Help accountability buddies break this barrier by making it a stipulation that buddies must ask for help on at least ONE thing each week! This might be help with a homework question, help feeling less nervous about an upcoming test, help making sure notes are complete by comparing with one another, or something else!

Motivate *Action 2: Make It Matter*

Why

When students feel as if what they're learning has value *to them*, they are more likely to feel motivated to put in the effort to learn it!

When

Any time you have the opportunity to connect a lesson with a skill or subject that matters to your students, do it!

How

1. Pay attention! My neighbor Jim Collis, who taught transferrable job skills to at-risk youth for 2 decades, once told me that the most effective way to motivate students was to figure out what mattered to them, to connect your lesson to that thing, and to build their self-esteem with regard to all of the above. As he put it, "Kids will tell you what matters to them—you just gotta be paying attention."

Observe your students when they talk to you or to one another. Keep notes about what matters to them: Are there certain celebs they like? Stores they shop at that you might base a word problem on? Goals they have that you might be able to relate a math-adjacent skill to? Jot down notes next to each student's name—this will come in handy later!

2. Make the connection! Use your notes to make connections to what matters to either specific students—or to your class as a whole—when teaching a lesson! You might introduce a concept like financial math by investigating how much money students' fave artists actually make via streaming their music on Spotify (*Spoiler*: My band made $7 last year.), or you might emphasize the math-adjacent skills that might come out of learning financial math, like the ability to budget, save, and spend!

IN THE MOMENT

Jim also told me that, in his experience, "at-risk" usually just means "haven't found their niche in life yet," and I thought that was a really nice way of looking at it. What do you think? What if we gave *our* students the grace and understanding that Jim gave his?

Grade-Level Modifications

- **Grades K–2:** Crowdsource! Ask your students what storyline they want to see in their next word problem and have them pitch in. You'll end up with some pretty ridiculous (and hilarious) problems, but your class will have A LOT of fun solving them!

- **Grades 3–5:** Something small like putting students' names and interests directly into word problems can go a long way at this age! Have fun with it and make sure every student has a chance to be featured so that no one feels left out.

- **Grades 6–8:** This is that weird age where some kids are watching adult content on TV while others are still keeping it PG, so keep that in mind when tailoring content to your audience!

- **Grades 9–12:** Feeling stuck? Ask students what the latest TikTok trend is. . . . and you'll be set! You can even use social media trends as alternate modalities of presenting assignments. For example, have kids create TikTok dances to express a math concept or use short-form video to teach a math lesson as concisely as possible!

Motivate *Action 3: Hit the "Reset" Button*

Why

Students have been taught to associate success in math class with high grades, period. While Step 3 is all about challenging that assumption, we can't expect them to suddenly forget all of the messages they have received up until this point—we just need to find effective ways to remind them that success is about more than marks! When they are reminded that they no longer have to adhere to archaic definitions of success, they have the freedom to be motivated by what matters to them! As we help students hit "refresh" on their old views of success, their new definitions replace the old, forming into long-lasting beliefs.

This action is all about cognitive restructuring (Stanborough, 2020), a technique where you do the following:

1. Identify what distortions you hold. *Example: If you don't get high marks, you suck at math.*

2. Explore why you hold those distortions. *Example: That's what I have been told my whole life.*

3. Investigate whether that distortion is harmful or not. *Example: It makes me feel like I'm bad at math if I don't get high marks, so I avoid even trying, so yes, this distortion is harmful.*

4. Cognitively restructure it. *Example: There are other things that make you good at math that don't involve grades. For example, if I work hard to solve a problem and don't get the right answer but I learn something in the process, that's one way to feel good about math!*

Over time, the repetition of hitting "refresh" turns outdated beliefs about success into new updated beliefs!

When

This is something you can do every lesson as you notice that old beliefs about success pop up among your students!

How

1. Actively listen! Be on the lookout for any statements from your students that might imply they feel unsuccessful because they're not getting high marks (or at least high relative to their expectations), not doing as well as someone else, not getting the right answer right away, or anything that implies an outdated version of success. When you hear it, take a pause to address it in a curious, open-minded way!

2. Unpack outdated definitions of success! Don't be all like, "OMG, YOU JUST SAID SOMETHING BAD" (of course I know you wouldn't do that!), but instead take an opportunity to address something you've just heard with the student who expressed it. You might say something like, "It sounds like you feel unsuccessful because you didn't get an A on this test, but does that *really* make you unsuccessful? Why do you think that?" Give them the space to express why, in that moment, they're defining success by their grades. Where did they get that idea, and do they truly believe it?

3. Hit refresh! Ask your students what they really think "success" means in math class and to compare it to those outdated definitions they may have walked into your classroom with. Compare and contrast, and then come up with new definitions of success that align with the values of a Math Therapy classroom. Make these definitions visible, every day, by creating classroom posters or charts for your walls. Consider reminding students of what success has been redefined as by starting each week with a celebration of student success. You might ask kids to share a "successful moment," and you might share some of your own to show the diverse ways in which success happens for everyone, every day! If you keep helping your students reaffirm their new definitions of success by hitting refresh, they will slowly start to wrap their minds around more diverse definitions of success that allow them to *feel* successful and that motivate them to keep building a better relationship with math!

Grade-Level Modifications

- **Grades K–2:** Model success every day by being super extra about pointing out your own "successes" as well as those of students so that they can start getting used to the diverse versions of success that exist within the classroom!

- **Grades 3–5:** Get one of those big, red reset buttons from the office supply store and place it on your desk—get one that makes a sound! When an outdated definition of success pops up, hit the reset button and get kids excited about resetting their definitions of success!

- **Grades 6–8:** When asking kids to consider newer definitions of success, prompt them by bringing up people they admire that may have dealt with a similar situation as they're going through right now. For example, they might say they feel unsuccessful because they get nervous and anxious every time they have a test. Bring up someone they admire, like Olivia Rodrigo. Do they think she's successful (duh, yes!)? Do they think that she gets nervous and stressed before a performance sometimes (definitely, yes!)? Well, then being nervous and stressed doesn't make you unsuccessful—bam bam BAM!

- **Grades 9–12:** You can spend a bit more time exploring where their definitions of success come from and how they might be either outdated or not serving them in a productive way. Focus on how they might modify their definitions of success so that that they are motivated—instead of discouraged—by them!

Motivate Activities

Use the following Activities to motivate your students on a deeper level!

Motivate *Activity 1: Redefine "Success"*

Why

When students take ownership over success and have the freedom to define it in a way that matters *to them* and leads to *authentic pride*, they are more likely to be motivated to achieve it!

This Activity is so, so important and is a total gamechanger when it comes to motivating your students to *want* to build a better relationship with math. Remember, as much as it sucks to hear it, many of our students feel that math class has no purpose and that the skills they're being told to build serve no meaning. When we open up the definition of "success" to include skills and markers that actually matter to our students, we see a shift in their motivation. AND when we open up the definition of "success" to allow for student voice and autonomy, we see a further shift in motivation!

When

I think we all assume that everyone has the exact same definition of success, so we don't bother talking about it. Start your next lesson with a discussion about success. Moving forward, find little opportunities to reinforce the idea that everyone's definition of success is valuable and worth striving for, that those definitions can change, and that you are there to help each student achieve success in a way that matters to them!

How

1. Get talking! (See the free download on the Math Therapy website.) Set 10 minutes aside at the beginning of your next lesson to open up a discussion about what success means. Here are some ideas to get across:

   - Success means different things to different people, and you are genuinely curious what success means to everyone in the room—there is no right answer!

- You might share the definitions of failure and success I shared earlier in this chapter and ask your students if they agree. Why or why not?

- Share that you've been giving this question a lot of thought and have come up with some definitions of success that are meaningful to you—share those with the class.

- As students come up with ideas, jot them down on the board. Consider organizing them into columns (see below). You might also break them down into columns labeled "External Validation" vs. "Internal Validation." The key is to get your students thinking about ways that success can be felt and measured that don't just require *someone else's* opinion (external validation).

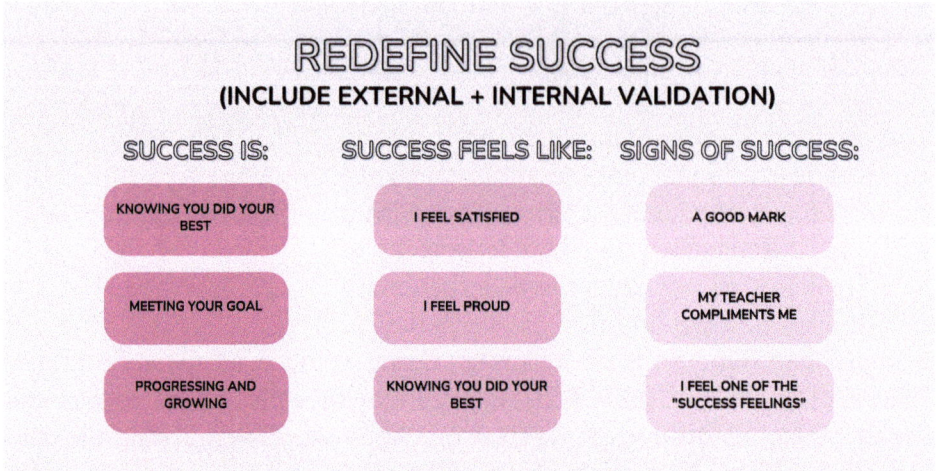

REDEFINE SUCCESS
(INCLUDE EXTERNAL + INTERNAL VALIDATION)

SUCCESS IS:	SUCCESS FEELS LIKE:	SIGNS OF SUCCESS:
KNOWING YOU DID YOUR BEST	I FEEL SATISFIED	A GOOD MARK
MEETING YOUR GOAL	I FEEL PROUD	MY TEACHER COMPLIMENTS ME
PROGRESSING AND GROWING	KNOWING YOU DID YOUR BEST	I FEEL ONE OF THE "SUCCESS FEELINGS"

♡ IN THE MOMENT

It's totally valid if students say that good grades and success go hand in hand because that's, like, the message they have been receiving their entire lives. The key is to emphasize that while good grades can be ONE side of success, they're not the be-all end-all of what it means to be successful! Think about how you might do that and reflect on whether or not you're reinforcing this message throughout your words and actions!

2. Make it math-y! Spend 5 to 10 minutes refocusing the discussion so that kids are sharing what success means *in math class*. A lot of what they shared in the last step will still apply!

3. Make it stick! It's time to put the results somewhere everyone can see them. If you're anything like the teachers I work with, you LOVE classroom posters. I love making downloadable poster packs for teachers, as I find that they're a nice visual way to reinforce the core ethos of a Math Therapy classroom—BUT, in this case, I want you to make your own. Make a giant poster containing all the definitions you come up with or a series of posters (one for each definition). If you don't have your own classroom, fill out the downloadable PDF provided on the Math Therapy website, print it out, and make sure your students place it at the beginning of their Math Therapy journals or whatever binder or notebook they use to do their actual math work in so that they can see it every day!

4. Reinforce the message! Every day, find little opportunities to reinforce the message that success is unique and valuable to each individual in order to keep them motivated! This might take the form of adding a reflection question at the end of an assessment that asks, "Name one successful thing you did on this assessment and tell me why you felt it was a success." It might mean making an effort to dish out compliments like, "I'm really proud of you for trying some of your questions on this test instead of leaving it totally blank—that's progress!" or "You may not have gotten the grade you wanted, but did you notice that you didn't freak out during today's assessment? That's a huge win!" Your compliments will depend on how your students have redefined success for themselves. Take note of who has said what because it tells you WHAT your students value and WHAT motivates them to keep going. This is the most valuable knowledge you can have when it comes to giving each of them the motivation to build a better relationship with math, even when it seems like the hardest thing on the planet to do!

 **Grade-Level Modifications**

- **Grades K–2 and Grades 3–5:** Here's my fave fun, DIY way to get those success definitions onto your wall: Create a giant cutout of a suitcase, call it a "Success Suitcase," and have them unpack it by creating bumper stickers with success definitions that will cover the suitcase! On our Math Therapy journey, we are doing A LOT of unpacking and carrying A LOT of lessons with us, so this suitcase is the perfect metaphor!

- **Grades 6–8:** Allow kids class time to each create a poster for one of the success definitions that resonates with them. Encourage them to create a visual representation (doodle, colors, etc.) to showcase what their chosen success definition looks like or feels like to them. Consider holding a success show-and-tell and allowing kids to share!

- **Grades 9–12:** Instead of making classroom posters, after your discussion has ended set a timer for 7 minutes, ask students to open their Math Therapy journals, and answer the prompts:
  - What does success mean to me?
  - What does success mean to me in math? (I would specifically omit the word "class" because this is about their relationship to MATH, not their relationship to your classroom—it's more open-ended this way.)

Collect your students' reflections so that you can learn more about what motivates each of them!

HOT TIP

Consider making this an ongoing discussion throughout the school year, revisiting and reflecting on the concept of success in different contexts (e.g., when taking a math test, when reviewing your report card, when helping a friend, etc.).

Motivate *Activity 2: In Your Dreams*

Why

When students are connected to their purpose, they are not only more motivated to put in the work to achieve their goals, but they are happier *in general*.

I love a vision board—who doesn't!? No matter what age you are, vision boarding brings back the nostalgia of flipping through magazines, making collages, and getting inspired by photos and artwork—that is . . . unless you're a teenager and have literally never seen a magazine because you live on TikTok! The good news is that scrapbooking is cool for teens, and Pinterest and Instagram are perfect for printouts that can be thrown on a vision board, so hey—you're in luck! This Activity is a relaxed way to have your students discover what matters to them and a really effective way for you to see that— visually! (See the free download on the Math Therapy website.)

When

Take 20 minutes to introduce the assignment, let students percolate over-night and print out anything they might like to add to their vision boards, and then give them the next lesson to create their boards!

How

1. Get inspired! Explain that a vision board is a piece of paper on which you put all your hopes and dreams. Vision boards can be as silly or serious as you want them to be, but at their core, they're simply a visual representation of stuff that brings you joy, that you either have now or want one day, or that just makes you happy to look at! Don't take it too seriously; the idea is to allow students to get in the flow of figuring out what makes them happy in a low-stakes way! Show them examples of vision boards (there are a ton on the internet) and give them a day or so to figure out if there are any materials they want to bring from home to put on their board. These might include photos, printouts, or cutouts from newspapers or magazines.

2. Cut and paste! On the day that you will all create your vision boards, make sure you have old magazines on hand in case students need some help. Make sure every student has access to an 8.5 × 11 piece of paper (cardstock is even better if you have it!), scissors, and glue or tape. Give them the entire lesson to create their vision board and allow them to take their boards home to put on the finishing touches if needed! Finally, ask that on a separate piece of paper they name three of the things they put on their board and explain what they mean to them.

3. Look closely! Once students hand their boards in, this is your opportunity to learn something new about each one of them. You might jot down some follow-up questions for each of them or go through your class list and make notes. You now have more valuable info on what matters to each of your students, which means you are even better equipped to motivate them! Not only that, but in making their boards, students were engaged in reflecting on what matters to them and what they're aspiring to in the future, which just feels good and puts them in—you guessed it—a motivated state!

 **Grade-Level Modifications**

- **Grades K–2:** Consider providing pre-cut shapes, stickers, and images printed on sticky paper for kids to use on their boards to avoid the mess of glue and cutting!

- **Grades 3–5:** Things can get messy! Make sure you use glue sticks instead of liquid glue and consider providing poster board instead of cardstock so that kids have a bigger surface to work with! Allow them to draw and doodle as well as cut and paste!

- **Grades 6–8:** Vision boards and collages can be so much fun and allow kids to really get creative and add emotion to their projects. Consider involving stickers, glitter, markers, and upcycled supplies so that kids can really make their visions come to life!

- **Grades 9–12:** Consider allowing students to make virtual vision boards by using online images. If you're strapped for time, you can have kids do this for homework and submit electronically.

 Motivate *Activity 3: My Mathspiration Is . . .*

Why

When students are able to connect math to something that matters to them, their motivation drastically increases. By having students learn more about people who actually do math, we not only diversify the definition of what "math" is, but who does it and why it matters!

This Activity invites students to find their mathspiration—a fictional or nonfictional character who uses math in a cool way. This person might be a superhero, reality show contestant, celebrity, animal, cartoon, video game character . . . the options are limitless! Your students will produce a one-page report that details what math this person uses, how, and why they find it inspiring! Inspiration fuels purpose and meaning, and remember what we said about states? When students are inspired, they are likely to feel motivated! (See the free download on the Math Therapy website.)

When

This assignment can be done individually, in pairs, or in randomized groups of three. I suggest dedicating one lesson to explain the Activity and to allow students to gather with their partner(s) to start brainstorming which fictional or nonfictional character they want to focus their project on. Students can then do their research and complete the assignment for homework!

How

1. Make math matter! Begin by reminding students that math is more than just adding numbers and cranking out answers and that most of the people they love and admire are doing really cool things with math. Share a few diverse examples, including superheroes with special skills (hello, Spiderman is legit ON TOP of his geometry game with those webs!), reality show contestants (don't you think the candidates on *The Bachelor* are fully calculating the odds when they hand out those roses?!), and even celebrities (let's talk about when Taylor Swift RE-RECORDED ALL HER SONGS because she didn't own rights to the original masters . . . fast-forward to her being an ACTUAL BILLIONAIRE NOW!). Encourage them to start sharing examples of their own and let a healthy debate ensue! They will likely start to debate whether the math skills being listed are "actual math," who should be included in this list, or whether the characters being talked about are inspiring or not. Chris Luzniak (2020) has taught me so much about the power of debate in math class and how it can foster collaboration, meaning, and purpose, so any time a mathematical debate ensues, I encourage teachers to let it happen! Remember, students are motivated by what's relevant to them RIGHT NOW, and nothing is more relevant than winning an argument!

2. Set expectations! Explain the goal of the project, which is to pick a fictional or nonfictional character and describe how they use math in a cool way. The output is simply a one-pager, so this isn't a huge assignment, but an impactful one!

3. Brainstorm! If you're going with groups, assign them and have everyone shuffle around so they're sitting with whoever they're working with. Hand out a template of the prompts they are being asked to consider and allow them time to discuss the person they might want to profile. Some extra prompts might include:

   a. Who did you choose to profile?

   b. What do you love about them?

c. How does your character use math? List the math skills they use and give concrete examples of scenarios in which they have used those skills!

d. Why do you find the way they use math inspiring?

Make sure you take a look at the character each group has chosen to profile so you don't end up with something inappropriate . . . I've heard some whack stories, let me tell you.

4. Share! Once the one-pagers have been completed, have students present their chosen mathspiration to the class! Ask questions to further the discussion around how math can be found everywhere and used in the most unexpected circumstances to motivate students to consider how this is true in their own lives.

Grade-Level Modifications

- **Grades K–2:** Give students a set of five questions for them to answer about their mathspiration instead of asking them to freewrite!

- **Grades 3–5:** Give kids a list of 10 to 15 diverse examples they can choose from if they're having trouble coming up with their own personality to profile!

- **Grades 6–8:** Spend extra time on the first part of this Activity to get students thinking about their favorite characters and people and the variety of ways in which they're successful! This will pay off when it comes time for them to choose someone to focus their project on.

- **Grades 9–12:** Allow students to create their profile in ways other than a one-pager! Give them the option of creating a TikTok, filming an interview where they role-play interviewing their chosen person, or creating a mini podcast!

 ## *MOTIVATE* TOOLKIT: SIMPLE SWAPS

You are doing an amazing job at cultivating a Math Therapy classroom, and I KNOW that your hard work is paying off. To add a little more *oomph* to your Math Therapy practice, here are some simple swaps you can make to enhance all the work you're already putting into Step 3. These substitutions may be simple, but they pack a punch. Sometimes it's the smallest things that make the biggest difference, so don't underestimate the effectiveness of the baby steps you're taking every day!

SWAP THIS	FOR THIS
Giving students a list of topics to choose from for a capstone project or assignment	Allow students to select topics, projects, or activities based on their interests
Reusing lesson plans without altering them	Make tiny but meaningful tweaks to your lesson plans that integrate students' personal interests!
Prioritizing individual work and individualized assessments	Encourage collaborative projects that involve students working together and consider trying out group assessments
Taking on all the emotional labor of inspiring students yourself	Bring in speakers or graduate students to talk about what their math journey taught them along the way and how they use those lessons in their line of work or field of study
Strictly paper-based assessments	Make room for assessments that use tech (making videos, recording math songs) or allow students to present their knowledge orally
Only talking about traditional mathematicians found in textbooks (Einstein, Pythagoras, Archimedes)	Talk about folks who are good at math that we don't normally hear about (Ariana Grande, Danica McKellar, Natalie Portman, John Urschel, Brian May . . . !)
Celebrating good marks	Celebrate success milestones that are meaningful to your students (e.g., improvement on a skill, progress in mitigating math anxiety, taking a risk, helping another student)

HOT TIP

Give *constructive* and *specific* feedback! Telling students exactly what it is they did well (e.g., "I love the way you used diagrams to try to solve these problems" vs. "great work!") builds their sense of self-efficacy and gives them concrete points of success that fuel their motivation!

MOTIVATE TOOLKIT: THEY SAY, YOU SAY!

It's hard to hear kids express that they just aren't feeling it during a lesson. Here are some of the common things we hear from our students and how you might respond!

WHEN THEY SAY . . .	YOU SAY . . .
"But when will I ever use this irl?"	"When you lift weights at the gym, you're not doing it because you're going to start trying to lift random heavy objects irl. You do it to build strength to do the things you love! The same is true here: You may not use SOHCAHTOA irl, but you will def use the skills you built while learning it—resilience, perseverance, inner strength—irl!"
"This is boring."	"I totally get that, and I know it sucks to feel like you have to do boring things. How could we make this more interesting for you?"
"What's the point if I can just use a calculator?"	"Don't think of it as calculating—think of it as THINKING! You're training your brain to think in a totally different way than it normally does day to day. Sure, you could use a calculator, but then your brain would miss out on the chance to develop this cool new way of looking at and solving problems!"
"I don't care."	"I have felt that way so many times, I get that. If you don't care about this, what DO you care about, and is there a way we can connect the two?"

 ## *MOTIVATE* TOOLKIT: EXPANSION PACK

Motivating students to genuinely WANT to build better relationships with math is one of my favorite parts of Math Therapy because it allows us to learn so much about our students and what drives them, but more than that—it encourages THEM to figure that out FOR THEMSELVES! Most students are never asked what matters to them, and this will be the first time that many of your students have been asked that by an adult—it's a really special time, and it's so important for them to know that WE *DO* CARE ABOUT WHAT MATTERS TO THEM! Math Therapy is about so much more than math class, and this step illustrates that. Yes, what matters to students helps us motivate them in math class, but what could be more important OUTSIDE the classroom than truly knowing what brings you purpose and pride in this life?! If it were up to me, this would be a whole class in school on its own, but since it's not up to me . . . here are more of my fave *Motivate* Activities that you can deep dive into if you have time. You can find templates and instructions for most of these Activities by heading to the online resources on the Math Therapy website!

1. **A Cloud of Success:** Use mind mapping as a tool for students to visually organize their thoughts about success! Ask them what success means to them and use a word cloud generator like www.freewordcloud generator.com to create a visual representation of what success means to the students in your classroom, which will not only get them feeling motivated but help YOU figure out how to best motivate them. You might go further by asking them to share words that come to mind when "success in math" is mentioned (as well as "success" in general or in other subjects) and prompt a discussion about the similarities and differences. The bonus of using this visual tool is that results are anonymous, and you can then ask students to expand on certain statements without calling attention to who actually wrote them!

2. **Set SMART Goals:** Use SMART goal-setting to help students turn their visions of success into tangible, reachable goals (see the free download on the Math Therapy website). When success turns into something tangible and reachable, students are more motivated to head in its direction because they actually have a direction to follow! SMART goals are

Specific: Clearly define what you want to achieve. Be specific about the goal, avoiding vague or general statements.

Measurable: Establish criteria to track progress and determine when the goal has been achieved. Use quantifiable metrics whenever possible.

Achievable: Ensure that the goal is realistic and attainable.

Relevant: Align the goal with your overall objectives and the broader context of your work or personal goals. This is the part that brings purpose to the pride!

Time-bound: Set a specific time frame for achieving the goal so that you can celebrate the achievement and set a NEW goal!

Here's an example of turning a general goal into a SMART goal:

- **General Goal:** I want to improve my relationship with math.

- **SMART Goal:** By the end of the month, I want to be able to write a math test without leaving half of it blank. To help me do that, I'm going to start studying a week before each test, ask my teacher for help if I don't understand a concept, and work on mindfulness techniques to help me relax on test days!

In this example, the SMART goal is more specific (improving the number of questions attempted on a test), measurable (more than half), achievable (through practice, teacher support, and mindfulness), relevant (to building a better relationship with math), and time-bound (by the end of the month).

3. **More-Than-Math Speaker Series:** Invite guest speakers from diverse professions to share their experiences and perspectives on success. Allow students to ask questions and engage in discussions about different pathways to success, not just those that involve math! The idea is to get students talking about what success really means to them and the diverse ways they can go about achieving it. As a bonus, students will get to mythbust ideas that success and failure are opposites and will likely learn that failure is a large part of everyone's success story!

4. **Dear Future Me:** Ask students to write a letter to their future selves, describing what success looks like to them and what steps they plan to take to achieve it. They can revisit these letters monthly or quarterly to check in to see if they're on the right track. Consider meeting with them one-on-one to ask them how they're doing with achieving their vision of success and what you can do to support them!

5. **My Personal Brand:** Assign a project where students develop a personal brand, including their values, skills, and goals. Talk about brands that students love to set the stage. What is the ethos of these brands? What are their core values? How are those reflected in the products they sell and the marketing they put out? This is a fun project that encourages self-reflection and helps students articulate their vision of success in a way that's relevant and reflective of their lived experience as consumers (of literally everything)!

6. **My Success Story:** Have students create a timeline of their past successes, both big and small, and to make note of what skills they used in achieving those successes. This visual representation can help them identify patterns and factors contributing to their successes and help them strategize a path forward to feel successful when it comes to their relationship with math! When students see that they have achieved so much and developed so many skills to help them along the way, it adds to their sense of pride and motivates them to keep going!

STEP 3 TOOLKIT
TO GO
motivate

motivate
ACTIONS

1. ASSIGN ACCOUNTABILITY BUDDIES

Assign each student an accountability buddy to keep th math positivity flowing!

2. MAKE IT MATTER

Find ways to show kids that the math-adjacent skills they're learning are valuable to them irl!

3. HIT THE "RESET" BUTTON

Help students reset the myths they hold around success so that math class is about more than marks.

motivate
ACTIVITIES

1. REDEFINE "SUCCESS"

Construct new definitions of what success means in your classroom.

2. IN YOUR DREAMS

Get your students to make vision boards to discover what matters to them.

3. MY MATHSPIRATION IS...

Encourage students to find a cool celeb or superhero that uses math in a unique way!

motivate
EXPANSION PACK

A CLOUD OF SUCCESS ✳ SET SMART GOALS ✳ MORE-THAN-MATH SPEAKER SERIES ✳ DEAR FUTURE ME ✳ MY PERSONAL BRAND ✳ MY SUCCESS STORY

BEFORE WE MOVE ON

You made it! You did it! Now, take a few moments to process, reflect, and treat yourself!

Treat Yourself

What brings joy and value to YOUR life? Hanging out with friends? Self-care? Sleep hygiene? Pick something to do for yourself that you know will add value, right now. Call a friend and make plans, get a mani-pedi, or take a nap: dealer's choice!

ASK YOURSELF

1. Did this chapter change or challenge the way you think about *Motivation*? Why or why not?

2. Have you labeled certain students as "unmotivated" in the past? Did this chapter challenge or change the way you view those students? Why or why not?

3. Which of your students do you think will benefit most from the Actions and Activities suggested in this chapter? Which do you think won't benefit as much? Why?

4. Which Action or Activity are you most excited to try with your students?

5. What is something in this chapter that makes you go "aHA, this is totally going to make a difference"? Why?

6. What is something in this chapter that makes you go, "That will NEVER work with my students!" Why?

7. What are some of the challenges you may face when trying to implement the suggestions in this chapter? How might you prepare for these challenges ahead of time?

$\oplus \heartsuit \pi$

CHAPTER 6

MATH THERAPY STEP 4

Makeover Math Identities

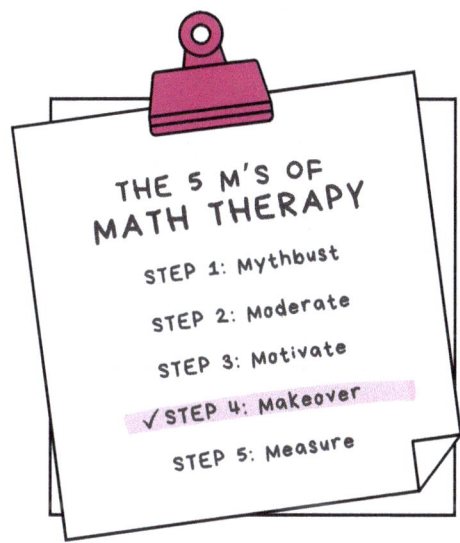

THE 5 M'S OF
MATH THERAPY

STEP 1: Mythbust

STEP 2: Moderate

STEP 3: Motivate

✓ STEP 4: Makeover

STEP 5: Measure

In this chapter, we get to:

1. Discover how our math stories came to be and how they affect us

2. Learn all about our brain's tendency to always think the worst and find out what we can do about it

3. Master techniques that will help our students shift their math stories and transform their ability to build better relationships with math

4. Dig into our Step 4 *Makeover* Toolkit and empower ourselves with transformative Actions and Activities to help students makeover their math stories

A couple of months ago, I tutored a Grade 10 student, Ella. She was working on linear systems, and when she came in, she was in tears because she hadn't understood anything that happened in class that day and swore up and down that she was NEVER going to "get it" because it was impossible. Not to brag, but within 15 minutes, I had her using substitution and elimination without a hitch, and, guys, she even had that moment we all dream of as teachers when she said, "OMG, this is SO EASY!"

I paused and was like, "Ella, can we just notice the fact that legit 15 minutes ago you walked in here CRYING and claiming you were NEVER going to get this, and now, you're actually so good at this that you could tutor ME in it?" You know what she said? She goes,

"Well, no. Now, I just feel dumb for not getting it earlier."

Wow. How could that POSSIBLY be her takeaway?! Here I was, marveling at the miracle that is neuroplasticity, and here she was, focusing on the fact that she must have been stupid for not understanding everything literally instantaneously.

Don't worry, this story has a happy ending because I obviously stopped our lesson right there, went to my supply cabinet, and brought out one of my favorite rose quartz crystals, handed it to her, and told her that I wanted her to hang onto it as a reminder to celebrate the fact that our beautiful brains can grow and change and that we must be compassionate and gentle with them. We had a chat about neuroplasticity (which she knew nothing about), and she got kind of excited about the whole thing. But wow, that was a close call.

This whole Ella situation has happened to me literally hundreds of times—and I'm sure something similar has happened to you, too! This is such a common narrative to hear from our students, and it illustrates the brain's negativity bias (Müller-Pinzler et al., 2019), which is our natural tendency toward thinking the worst (we'll explore that in more depth later!). As I discovered through our continued time together, Ella's math story was riddled with experiences and events that had convinced her that she wasn't "naturally good at math," and once we started to give that story a makeover, the way she approached math started to change.

STEP 4: *MAKEOVER!*

 ### IN THE MOMENT

Pulse check! Remember the mathography I had you write at the beginning of this book? Now is a good time to pull that up so you have it handy as you read this chapter—you'll want it close by!

Remember when you wrote your mathography at the beginning of this book, and I told you that we all have stories we tell ourselves? Well, we're about to revisit those stories, and just like I promised, you're going to have the chance to revise your mathography!

In this step, you will help your students discover what their subconscious stories are and help them rewrite those stories so that they reflect the empowering and limitless nature of their reality! But you won't stop there. Once those stories have been rewritten, you are going to help your students actualize them into being so that the lived experiences of your students begin to reflect their new narratives.

MAKEOVER

The fourth step of Math Therapy is all about empowering our students to rewrite the negative math stories that have held them back from believing that they can build better relationships with math. In this step, you will discover how to help students recognize what these stories are, where they came from, and how to give them a *Makeover* by rewriting new, positive math stories to replace the negative ones!

Why It Works: This powerful step equips your students to take everything they've learned from their Math Therapy journey so far and use it to take a major step toward rewiring their brains. By taking both *Mythbusting* and *Moderating* to the next level, students will formally rewrite the scripts that have been running on repeat in their subconscious. This step works because it is the weight of the storyline of our past math traumas that prevents us from moving *forward*. It works because if we *don't* rewrite the scripts that we live by we will inadvertently keep repeating the past. It works because the *Makeover* Toolkit transforms valuable research about the science of self-talk into practical strategies you can easily integrate as part of your current classroom practice. When students put pen to paper to rewrite their math stories, they give themselves permission to start a brand new chapter in their relationship with math!

How It Works: The Step 4 *Makeover* Toolkit is full of strategies you will use to rewrite the math stories that permeate your classroom so that the vibe goes from horror-movie-dramedy to feel-good-rom-com! It will also equip your students with Actions and Activities that get them closer to discovering what their "old" math stories are, recognizing how those stories affect them, and realizing they have the power to change them. In tandem, *your* deeper understanding of what they believe about themselves coupled with *their* readiness to open a new chapter in their relationship with math turn Step 4 into a double whammy!

I've said it before and I'll say it again: None of the 5 steps of Math Therapy are just one-and-done, but instead they are part of a set of ideas and strategies that will help you *build* and *maintain* a Math Therapy classroom. Some of the tools in your toolkit will be things that you can do *once* with your students, and others will need to be used *consistently* and *repeatedly*. The only difference in this step is that there is ONE key Activity that I really, REALLY want you to do with your students, no matter what, and it's the mathography Activity I asked YOU to do at the beginning of this book, remember? It's all about rewriting your math story, and it is POWERFUL. When we get to the *Makeover* Toolkit, I'll put it right up front so you know what I'm talking about. I just wanted to get this out of the way now so that you're emotionally prepared when we get there. Okay, moving on!

Now, before we get to the juicy contents of the toolkit, I think it's really important for us to understand exactly what I mean when I talk about math stories, where it is these stories even come from, how they actually affect our ability to build better relationships with math, and how we can use that knowledge to rewrite our stories to get closer to that fairy-tale ending we all dream of (or maybe that's just me, I LOVE a rom-com ending, ALWAYS)!

HOT TIP

Set realistic expectations! Explain to students that they have likely been thinking the *same* thoughts *for years*! They aren't just going to change overnight. With practice, repetition, and consistency, things change, but patience is key. Dr. Paul Conti (Robbins, 2023a) suggests that 4 to 6 months is a reasonable time frame within which to expect notable changes to our self-talk, but that doesn't mean that little changes won't happen along the way. Encourage students to enjoy the journey and notice the little wins instead of just focusing on the destination!

HOT TIP

Celebrate uniqueness! Always make it clear that our math stories are unique—just like all the stories we read about our favorite characters. There's no need to try to copy someone else's journey but instead have fun writing our own. Always try to be the best version of yourself, not a mediocre version of someone else!

MATH IDENTITIES: THE GOOD, THE BAD, THE UGLY

When I published my master's thesis in 2010, I argued that one of the reasons that many teenage girls didn't feel like "math people" was that there was a mismatch between *their* identity and the identity they expect to have *when they're doing math* (Vakharia, 2010). So all of the Grade 10 girls I interviewed said that being social and connecting with peers was high on their priority list—yet in most 2010 math classes, we were telling kids that math was a solo sport, and in most mainstream media, we were being told that to be a mathematician you must have zero social skills and be unable to hold down a romantic relationship (hello, *A Beautiful Mind* and *Mean Girls* and fast-forward to *The Queen's Gambit*, which came out in 2020 yet STILL reinforced the exact same stereotype A DECADE LATER!). For that reason, my research showed that many girls felt an incongruence between who *they were* and who "math" *wanted them to be.*

Math identity basically refers to how you see yourself in the math world. It's not just about the numbers; it encompasses our beliefs about our abilities, attitudes toward math, and our sense of belonging in the mathematical community. A positive math identity means that we feel like we belong in the world of math and believe that we can tackle math challenges,

> *My research showed that many girls felt an incongruence between who they were and who "math" wanted them to be.*

grow, and improve. On the flip side, a not-so-great math identity might make you feel—well—the opposite!

Math identity is shaped by various factors, including past experiences with math, societal influences, cultural background, and our learning environments. But what we don't often talk about is the fact that our math stories are an *inseparable* part of our math identities. Not only do our math

stories *shape* our math identities, but conversely, our math identities reinforce the narratives embedded IN those stories! And THAT is why it is IMPERATIVE that we get serious about identifying the etymology of our math stories and reconstructing them so that they propel us forward instead of holding us back!

 **IN THE MOMENT**

Have you ever thought about your math identity before? What influenced your relationship with math? Try to think about specific experiences, events, people, or anything else that played a role in shaping the way you feel about math!

What Even IS a Math Story?

As we have established by now, we all have stories that we tell ourselves, but often, those stories are built around our insecurities due to our brain's negativity bias (which we will talk about in the following section!). Our math stories are no different! They're essentially stories we tell ourselves about our relationship with math! I asked you to think about your math story at the beginning of this book by asking you to write out your mathography, and I'm going to throw a shortened version of mine below to refresh your memory as to what a math story might look like. I'm going to show you two versions: the one I wrote *before* I went through my own Math Therapy journey and the one I wrote after. Bookmark this page because throughout this chapter we're going to be revisiting these examples a lot!

My Math Story (Before Math Therapy)

*I failed math twice in high school because **I just wasn't a math person**. I'm more of the kind of person who likes the arts, so **it's harder for people like me to do math**. There were a few times that I had a tutor and actually understood the concepts, **but that's probably because they were easy concepts to begin with and I*** worked really hard at understanding them, **not because I'm actually smart** when it comes to math.

My Math Story (After Math Therapy)

I failed math twice in high school because **I believed I wasn't a math person.** In fact, most of my teachers didn't try to convince me otherwise, and while I did succeed at points when I had a tutor, **those victories were never celebrated.** When I worked hard, I was actually good at math, **proving that I AM capable of math! In fact, being artsy and creative certainly aren't a barrier** to math—they're additional skills I have that can be used to help me find creative solutions to math problems that other people might not think of!

IN THE MOMENT

Can you spot the differences? What changed between the first version of my story and the second? How do you imagine the first version makes me feel about myself vs. the second? Consider this throughout the chapter as we talk about how to help kids rewrite their math stories!

See what I did there? You'll notice that none of the FACTS changed, but my perspective did. My new story is a lot kinder to myself and puts me in the seat of power when it comes to my relationship with math.

Our math stories might be long or short, happy or sad, confusing or straightforward—there are no rules to the stories we've written about our relationship with math! However, for those of us who have rocky relationships with math, there are some commonalities that are worth highlighting, and those usually stem from how we "wrote" those stories in the first place. Once we understand how our stories came to be, that's when we can empower ourselves to impactfully rewrite them.

HOT TIP

If one of your students has an optimistic and positive math story that doesn't need a makeover, celebrate it and place emphasis on the parts of it that showcase growth rather than fixed ability. The more you seize opportunities to draw students' attention to their role in fostering a positive math identity, the more you empower them to moderate any future adverse experiences with math!

Where Do Our Stories Come From Anyway?

Most of our stories are built slowly, over time, beginning in childhood, and often, we don't update those stories with new, relevant information—meaning, we often get stuck in a loop, replaying the same story in our minds, whether it continues to be true or not.

Like, when I was younger, I fainted. It was a really hot day, and I was standing on an enclosed balcony, and there wasn't a lot of airflow. Maybe I was dehydrated? I can't remember. All I remember is that I fainted. It was the one and only time I have EVER fainted in my life. Since then, I have regularly done hot yoga, traveled to tropical places, been in hot, stuffy elevators—you name it—and I have been perfectly fine—except for the fact that as soon as I enter a hot yoga room, step into a tropical landscape, or enter an elevator, my entire body tenses up and relives the story I have created, which goes along the lines of "I am someone who might faint at any moment if I'm overheated." Despite the hundreds of experiences and pieces of evidence to the contrary, to THIS DAY, I still have this fear that if I get too hot I'm going to faint. Actually, wow, writing about this is totally therapeutic because I've never actually thought about this before. I have just accepted that when I enter a "hot" circumstance, I start getting anxious that I'm going to faint, my body tenses up, and quite frankly I FEEL faint until I use my mindfulness strategies (hello, Step 2!!!) to moderate my trauma response, breathe regularly, and feel "normal" again. But like, let's pause. How is it that despite ALL of this evidence to the contrary, I still carry around this story that I'm likely to faint if it's hot, and my mind and body react TO that story in real time, EVERY time? That is WILD. Wow. I'm having a moment with you guys right now.

Dr. Conti (Robbins, 2023a) explains that when we learn lessons in childhood we often *do not* test them again; we don't update those lessons with new information—we just accept them as fact and move on, integrating them as a central part of our storyline. Unfortunately, many of these storylines are not only untrue, but detrimental to us because they're built using false assumptions and erroneous correlations.

Let's take a look at some of the building blocks of our math stories so we can understand how to work with them productively instead of destructively!

Negativity Bias

I was born and raised in Toronto—aka the "citiest" place of all cities (at least in Canada). Now, in 2020 I decided that I wanted to live in the countryside part-time because I just like the peace and quiet and chill vibes, so I rented a house "up north" (aka north of Toronto). One morning, I woke up

HOT TIP

Don't be judgy! When students share their negative self-talk with you or with the class, try not to say things like, "OMG, THAT'S A HORRIBLE THING TO THINK!" Instead, ask them how that thought is making them feel or whether that thought is helping them or holding them back. Through gentle questioning and encouragement, let students know that thoughts aren't facts and that we can work together to change them!

during an intensely beautiful snowfall and discovered footsteps . . . in the snow . . . ALL AROUND MY HOUSE (see Figures 6.1, 6.2, and 6.3). I thought that was a bit strange because, like, why would someone be walking around my house unless they were, like, casing the joint or PLOTTING TO MURDER ME??!?

FIGURE 6.1 THE GAS METER CAPER OF JANUARY 20, 2021

FIGURE 6.2 THE PHOTOS I TOOK OF FOOTSTEPS IN THE SNOW

FIGURE 6.3 COME ON: IF YOU WERE ME, WOULDN'T YOU HAVE BEEN CREEPED OUT?

I decided to play detective and see if maybe other houses in my new neighborhood had similar murderer tracks around their properties. Well, let me tell you: As I walked the neighborhood, I noticed that, yes, THIS PATTERN WAS REPEATING! Every single house I passed had a similar trail of footsteps in the snow leading from the street, onto the lawn, and then AROUND the house. I started freaking out and snapping photos to build my case. The snow was rapidly falling, and I was afraid this EVIDENCE would soon be invisible! It was my DUTY to solve the case of the footsteps-in-the-snow caper, and I would NOT disappoint. I got home, shaken and panicked, and texted Cheryl, the woman whose house I was renting. Nothing. I decided I needed to take action before the Hallmark-worthy snowfall covered all the tracks, so I called the police. Now, guys—like, I have never called the police, ever. But this seemed like a no-brainer: There was a CRIMINAL ON THE LOOSE, and they were casing the neighborhood! The dispatcher heard me out and was super nice and sent an officer on his way.

I would like to point out that this was December, right around Christmas, and they sent an officer named BUCKY—like, talk about a holiday rom-com in the making! Anyway, so Bucky shows up and is super understanding and, like, "Okay, right, uh huh. . . ." And while he's jotting his notes down, my phone starts ringing off the hook, and it's Cheryl. So I pick up the phone, and Cheryl is like, "Vanessa . . . PLEASE tell me you DID NOT CALL THE POLICE." And I'm like, "I did, I'm on it, don't worry YOUR HOUSE IS SAFE!" Silence. And then Cheryl lets out a cackle UNLIKE anything I have EVER heard in my life. Like, the woman is OUT OF BREATH, laughing. And I'm puzzled, Bucky is on the doorstep looking at me like I'm crazy, and Cheryl finally catches her breath and goes,

THE FOOTSTEPS BELONG TO THE GAS METER GUY. THAT'S WHY YOU'RE SEEING THEM AROUND EVERY HOUSE. THE GAS METER IS LOCATED AT THE BACK OF EVERY HOUSE, AND ONCE A MONTH, THE GAS METER GUY MAKES HIS ROUNDS TO GO CHECK THEM ALL.

I. Was. So. Embarrassed.

Up north, they have a term for people like me. That term is *cidiot*. It's a mix between "city" and "idiot" and kind of refers to how "city-ish" Toronto people can be. It's not a nice term, obviously, and it's not my fault that we don't have this gas meter system in Toronto, but I DEFINITELY felt like a cidiot in that moment! Bucky was super nice about it and thanked me for "watching out for the neighborhood," lol, and that was our rom-com ending.

So what was happening for me in that moment? Why did I jump to the conclusion that I was about to be murdered? Well, this all ties back to our brain's **negativity bias** (Müller-Pinzler et al., 2019)! Our negativity bias is basically our tendency to focus more on negative experiences, feedback, or information, than positive as a mechanism for *keeping us safe*. In fact, I recently learned that back in the day, watching for FOOTSTEPS IN UNUSUAL PLACES is one of the things our ancestors evolved to be on high alert for because they needed to be hyper-aware of potential dangers like wild animals, predators, or enemies! Fast forward to today, and our brains still have this natural inclination to pay more attention to negative stuff. It's like a mental survival strategy that doesn't always serve us well in our modern, less life-threatening world!

The point is that, as a result of our natural inclination toward negativity, our math stories are often filled with storylines that put us on high alert. So for example, just like I had it in my head that mysterious footsteps around my house must be a sign of impending danger, someone might have it in their head that when there's a risk of getting a math problem wrong, they're likely to experience shame, panic, and confusion. The negativity bias makes them overly concerned about making mistakes, potentially hindering their willingness to take risks in learning.

Our negativity bias means that we recall insults better than praise, respond more strongly to negative events than to positive ones, and that we remember traumatic experiences better than positive ones (Cherry, 2023). Just think about Ella from my story at the beginning of this chapter! Instead of focusing on the fact that within 15 minutes she had learned a concept she thought she would never understand, she went right to focusing on the fact that there must be something wrong with her because she didn't get it SOONER!

> **Our negativity bias means that we recall insults better than praise, respond more strongly to negative events than to positive ones, and that we remember traumatic experiences better than positive ones.**

♡ IN THE MOMENT

Try everything! There are many strategies to try on the quest to changing negative storylines. I could write a whole book JUST about that! Here are some things you might want to try yourself if you're working on your own self talk:

- Investigate your self-talk by journaling.

- Try therapy.

- Share your thoughts with a friend.

- Read books or listen to podcasts that talk about the science of self-talk.

HOT TIP

The way to work through our negativity bias is by recognizing how it affects our math stories and then consciously choosing to rewrite our storylines so that they focus on the positive instead of the negative. As we learned in Step 1 through *Mythbusting*, we can train our brains to see the good stuff, appreciate the wins, and build a more positive mindset around learning, challenges, and life in general, and our Step 4 Toolkit will be full of strategies to use these skills to finally flip the script on our negativity bias!

Main Character Syndrome

Has the thought, OMG, I am for sure CURSED—bad stuff ALWAYS happens to me, ever crossed your mind? I'm not judging, just wondering. I legit have friends who think this (I can think of at least three of them). The funny thing is, if I turned it around on them and asked them if anyone else they knew might be "cursed," I BET they would tell me I was being ridiculous!

I call this "main character syndrome." It's basically when you decide that you're the main character in your own life movie and that you're extra special . . . but not in a good way. In fact, Dr. Conti explains that it is our tendency to see ourselves as "special" in ways that action our negativity biases into probabilities that aren't actually all that probable (Robbins, 2023a). There is an actual scientific term for this, and it's called the **self-effacing bias** (Herndon, 2018). This bias involves a tendency to attribute success to external factors and attribute setbacks to internal factors. For example, let's imagine a student who believes that they suck at math:

Success Attribution (External)

Scenario: This student scores a high grade on a math assignment.

Attribution: "Oh, I got lucky with this assignment. The questions were easy—believe me, it's not because I'm good at math—it's just a coincidence!"

Failure Attribution (Internal)

Scenario: This student scores a low grade on a math assignment.

Attribution: "OMG, I always mess up in math. I'm not smart enough—I'm just not a math person—you see, I was born this way!"

Notice that in this example the student attributes success to external factors like luck and downplays the role of their own skills in their success. On the other hand, they attribute failure to internal factors, emphasizing a perceived lack of ability rather than considering other possible reasons for the outcome. This pattern of attributions reflects self-serving bias. It prevents cognitive dissonance that reinforces this student's "math story," maintaining a narrative that externalizes success and internalizes failure.

As we help students recognize when and how they are doing this subconsciously (by retelling their math stories), we help them take control and rewrite their narratives in a way that reinforces that they are in the driver's seat when it comes to building a better relationship with math!

IN THE MOMENT

Your turn! Have you heard of the self-effacing bias before? If not, does it surprise you? Can you think of areas of life where it applies? Is your mind blown?

IN THE MOMENT

Cognitive dissonance is a doozy! This is the discomfort that arises when our actions and beliefs don't align, and it's enough to make us want to change our beliefs . . . or alter our behavior so that it aligns with existing beliefs. If we deeply believe that we are "not a math person," we're not going to be motivated to behave in a way that a "math person" would. We may not see the point in practicing or challenging ourselves to learn new concepts, for example. We can help our students relieve the tension caused by cognitive dissonance by helping them challenge deep-seated beliefs and rewrite their narratives AND change their behaviors so that the two align in a way that's beneficial to them!

Call Me, Maybe?

Have you ever been left on "Read" by someone you have a crush on? Like, you text them, you SEE that they've read your text, and then they totally ghost you and don't reply?! Because that JUST happened to me last week, and I was actually shocked at what happened next. Within a matter of hours, I watched myself go from "Cool, no biggie, he'll prob respond in a second," to "Weird that it's been an hour, but he's probably just busy," to "WTF?! OMG, did I do something?! Is he not into me anymore?!" to "WHAT AN A$$HOLE I CAN DO BETTER HE DOESN'T DESERVE ME!" Like, within 4 hours, I had gone from having a crush on this person to deciding that their lack of response clearly meant that they hated me, which meant I was going to hate them back. Four hours later, they texted me back to be, like, "OMG, sorry, the second I got your text, my flight took off, and I just landed!" That's right. The whole thing had been *in my head*. I had fabricated an ENTIRE story and, even worse, had convinced myself of its validity AND had both physically and emotionally reacted as though my FICTION WAS FACT. It was wild. But it's reflective of what most of us do MOST of the time; don't lie to me—I know you've done this too!

 IN THE MOMENT

Has your mindset ever affected your physical behavior? I have discovered that I can "talk myself out of" being able to sing high notes. It's actually wild, but there are times when I'm working with my vocal teacher and she's like, "I know you can hit a G, and you're talking yourself out of hitting an E" and she's right. When I get "in my head" about not being able to sing super high, it's like my throat closes up, and I can't do it even though I KNOW I can! Has anything similar ever happened to you?!

According to the National Science Foundation, the average person has about 60,000 to 80,000 thoughts per day. Of those, 80% are negative and 90% are exactly the same repetitive thoughts as the day before (Simone, 2017). Now, for people who have a negative relationship with math and a lifetime of math microtraumas, their math stories comprise a similar makeup of thoughts, and those are the stories circulating in their minds day in and day out. Just imagine you had someone following you around all day long, whispering, "You suck at math, you're a  moron, you always make mistakes, you're not good enough," and so on in your ear ALL day long. How would you start to feel about yourself?! THAT is what many of our students are going through, and THAT is why it's so imperative for us to help them recognize and take control over their math stories!

As you saw earlier in this chapter, our math stories may range in emotion and length, but regardless of what exactly it is that we're telling ourselves on repeat, there are three ways in which our stories undeniably affect our lived reality if we don't take control over their authorship. These stories trick us into feeling like we are

- passively participating in our stories rather than actively writing them,
- adding unnecessary heaviness to our forward momentum, and
- scripting out a self-fulfilling prophecy.

Left uncovered, our stories have the potential to drag us down instead of lift us up.

Things Don't Change If Things Don't Change

So I have this friend, Stacey, who is CONVINCED that she's "not a relationship person" and "totally undatable." Like, you know how Taylor Swift is all, "I'm the problem, it's me" (Swift, 2022)? That's Stacey's "relationship story." Her evidence? That every time she dates a guy, as soon as it starts getting serious-ish, they ghost her. Now, remember what I said earlier about external vs. internal attribution? This totally plays a role here. Stacey attributes all of her relationship failures to some internal thing that must be wrong with her, when actually her relationships have been doomed from the start for one KEY reason: She keeps dating guys who *do not want to be in a serious, committed relationship!*

In Stacey's mind, she's undatable and her 12 past failed attempts at dating prove that, but the thing is, it's not like those past 12 relationships ALL failed for different reasons. They all failed for the SAME exact reason, so really, it's like Stacey was simply repeating the same pattern over and over again, and guess what? She kept getting the SAME result!

If we want different outcomes, we have to change our behavior, period. We decided the next time she opened her dating app, I would be the one to swipe right for her because clearly she was choosing the wrong guys (aka the ones whose bios were tagged "not interested in a serious relationship")! I'm disappointed to say that those dating apps are full of red flags, and I had no luck, BUT a few weeks after our conversation, Stacey met a guy at the gym who was actually looking for a long-term, committed relationship. Guess what, guys? It has been over a year, and Stacey is in the relationship of her dreams. She changed her storyline (she stopped with the "I'm undatable" nonsense), she changed her behavior as a

result (she went after a guy who was actually emotionally available), and she changed her result (at the time of this writing, she hasn't been ghosted again)!

This is literally just like that student who's all, "No matter how hard I study for a test, I can't remember how to do a single question," but as it turns out, their version of studying for a test is reading their notes over and over and never actually trying to solve any of the problems. Their storyline tells them that studying just doesn't work for them, but the facts paint a picture of someone who has been doing the same thing (the wrong thing for *them*!) over and over again, expecting different results!

When we see our stories as unchangeable, we have less agency to rewrite them and are more likely to think of our abilities as fixed vs. flexible. When we see our stories as changeable and within our control, we are more motivated to alter our future behavior to get new results!

HOT TIP

Be nice! One simple strategy to start making inroads into changing our negative self-talk is to do something nice for ourselves or someone else. Encourage students to compliment one another or engage in an act of kindness every lesson in order to strengthen their positive thoughts about themselves!

A 20-kg Lie

By now, you might be thinking, "Ugh, *I* don't want to actually dig into the stories I've been telling myself and where they come from—why would my students?" and I totally feel you. No one wants to do this, it's not just you!

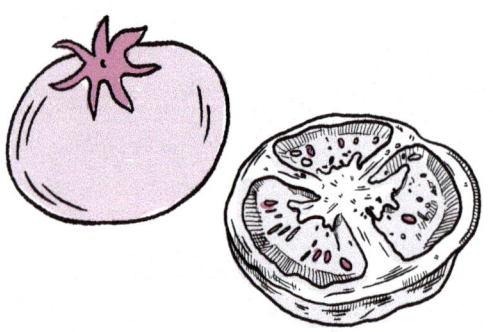

However, I promise that the more we hide from our past math trauma, the more it grows and festers and becomes an embedded part of our storyline. What's more, I think we can all agree in general that the more we try to ignore something or push it away, the stronger and more salient it becomes (what we resist, persists)! Seriously, I'm going to prove it—ready? Here's your challenge:

I dare you to NOT think about tomatoes for 3 whole minutes. I want you to set a timer right now, and ANY TIME you think about the word "tomato," or think about eating, smelling, seeing, or touching a tomato, or even imagine hot, juicy tomato sauce on a pizza . . . you have to reset the timer. Ready . . . set . . . go!

How did you do? If you're like most people I've tried this with, they reset the timer, like, five times and then just gave up on the exercise and begrudgingly told me I was right. Totally cool if you did the same!

♡ **IN THE MOMENT**

Think of a time you tried to hide from a thought. Did it work? Or did the thought come back to bite you in the butt? Asking for a friend. . . .

Since ignoring our math stories or the negative thoughts that make them up doesn't work, that means that everywhere we go we are *carrying the weight of those storylines with us*.

Remember, this doesn't necessarily even happen *consciously*, but that doesn't mean the weight isn't palpable. You know those hard-core runners that strap weights to their ankles just to, like, make running harder for themselves? Just imagine that was you and that you were dragging 20-kg ankle weights around everywhere you went—how would you feel? You would still be able to move forward, but that weight would undoubtedly make it much, much harder. That's what it's like when we carry our storylines along with us on the journey of building a better relationship with math. We don't know why, but it just seems HARD and TIRING and like we're taking one step forward and two steps back—because we are! We're constantly being dragged backward by the weight of our stories, and the only way to free ourselves is to face them head on and REMOVE them from our metaphorical ankles!

It's Britney, B*tch!

I saved the best for last because (a) it involves Britney Spears and (b) as educators, this is the most impactful and insidious consequence of *not* empowering our students to examine their math stories.

If you're a Britney fan, you know that Britney was a small-town gal from Louisiana with a big heart and even bigger dreams. Her breakthrough took place in the 1990s, an era where manufactured pop stars were all the rage (think: the Spice Girls, Backstreet Boys, and Christina Aguilera), and music labels would literally take a person and *mold* them into a pop-personality and that person would be expected to live, breathe, and sleep that personality at all times, especially when in the public eye. Now, Britney Spears—as we knew her then—was a *character*. Remember, the actual Britney Spears was a soft-spoken, mild-mannered girl who dressed conservatively, while pop star Britney Spears was a badass queen who sang lyrics like "It's Britney, Bi#ch" and wore show-stopping outfits. Now, before Britney actually got famous, she talks about how she would just have to "step into" her role as pop star–Britney, get on random stages in shopping malls where no one knew who she was, and act the part. And it was this continuous, repetitive act of *behaving* in line with her new pop star–Brit-Brit storyline that ultimately led to her building the confidence to *truly become* the Britney we came to know and love! Are you seeing those B-words guys? Believe? Behave? Become? (BRITNEY)?!

When you deeply hold a belief, expectation, or storyline, it can influence your behavior in a way that often brings about the fulfillment of that belief, expectation, or storyline!

Britney becoming *Britney* is an example of how one's storyline can lead to a self-fulfilling prophecy. When you deeply hold a belief, expectation, or storyline, it can influence your behavior in a way that often brings about the fulfillment of that belief, expectation, or storyline!

Just think about a student whose math story is riddled with statements like "Math doesn't come naturally to me" or "It takes me longer than everyone else to understand a math concept." If they carry that storyline into their math class, they might approach it with less enthusiasm, avoid seeking help, and maybe not put in as much effort. Guess what that will likely lead to? They're probably going to continue to struggle in math class, and suddenly, that storyline becomes a bit of a self-fulfilling prophecy. The belief influenced their actions, and the outcome aligned with what they expected. It's like a powerful loop of perception and reality, one that we MUST help our students break if we want them to truly build a better relationship with math. If human-Britney can become pop star–Britney through re-creating and reliving her new storyline, then I promise you that your students can go from where they are now to where you believe they can be by re-creating and reliving theirs!

IN THE MOMENT

Watch out! We're talking a lot about language and self-talk, and I just want to be clear that, as with everything, this CAN go too far. Just as I mentioned when we talked about growth mindset, language IS important in defining our reality, but it does not negate the real, lived consequences of systemic issues and oppressive systems. Make sure to be gentle when challenging your students' math stories, to acknowledge their lived reality, and to nudge them toward a better future!

PUT *MAKING OVER* INTO PRACTICE

It's time to give our math stories a makeover, and there's no way around it: We have to physically, actively *revisit* and *rewrite* them. I know that everyone rolls their eyes when "writing" is mentioned in math class, but as of late, there has been a push to increase proficiency in language and literacy in mathematics as we increasingly recognize how intertwined the two are (Noonoo, 2023), so think of this toolkit as a step toward doing just that!

This toolkit was built using the core principles of **narrative therapy** (Ackerman, 2019), an approach that focuses on the stories individuals tell about their lives. It recognizes that people construct their identities and make meaning through the narratives they create, which is why narrative therapy techniques are aimed at helping individuals reframe and reconstruct their stories to create a more empowering and positive self-narrative. In other words, narrative therapy is ALL about giving your story a mega makeover! Ready, set, GLOW-UP!

HOT TIP

Every student's math story is different, and some of them won't actually contain much negative self-talk at all, which is great. However, you will find that many students with positive math identities attribute their success to luck or being naturally gifted. Take the opportunity to encourage them to revisit their stories in a way that puts them in the driver's seat and allows them to celebrate their journey as being a result of THEIR hard work, not a result of chance!

YOUR *MAKEOVER* TOOLKIT

I know that I sound like a broken record, reminding you that each toolkit contains *Actions* and *Activities*, but I'm actually about to say something slightly different, so don't skip this part! The ONE difference between your *Makeover* Toolkit and ALL the other toolkits is that there is ONE Activity I need you to do, no matter what. I know it's not cool for me to play favorites, but when I built this step of Math Therapy, the most powerful thing to come out of it was the Mathography Makeover Activity. I'm telling you, it is so, so powerful—and that's

why I sneakily had you do the first part of it at the beginning of the book—remember that? If you don't do ANYTHING else in this toolkit, please consider doing that ONE Activity. You can find it on page 1, and honestly, it WILL make the BIGGEST difference!

Makeover Actions

Use the following Actions as a part of your regular classroom practice to help students makeover their math stories!

Makeover *Action 1: Choose Again!*

Why

By showing students how to reframe negative self-talk, we slowly help them makeover their math story piece by piece! We've spoken about the power of thought in defining reality and, additionally, the power of repetition in reaffirming our stories. By encouraging students to choose their thoughts carefully, we can help students rewire the way they think of themselves as math learners!

I learned this Action from Gabby Bernstein (2019) and love that it makes use of **cognitive restructuring,** a therapeutic technique used to help individuals recognize and change negative thought patterns. The process involves identifying negative thoughts; critically examining the evidence supporting them; generating alternative, more balanced thoughts; and testing these thoughts in real-life situations. When we "choose again," that's EXACTLY what we're doing! By repeating this process, students gradually replace self-defeating thoughts with more realistic and adaptive ones, leading to positive changes in emotions and behaviors, and ultimately, to mega math story glow-ups!

When

As you have already started to build a Math Therapy classroom in which students are becoming familiar with growth mindset language and the power of our thoughts, you can jump into this Action consistently and repeatedly, as needed, without preamble! Any time you hear a student say something that might benefit from choosing again, jump in with this Action!

How

1. Whenever a student expresses a thought that suggests that the negative aspects of their math story might be taking over, this is your chance to help them Choose Again! You can do this one-on-one with the student who expressed the thought, you can pause the class for a moment to

get the class involved (you read the room on this one!), and you can encourage students to do this for YOU if you're the one expressing the thought! For example, let's say students are all working away to solve a problem and you hear one of them mumble, "I suck at math. I'll never understand it." Head over to chat with that student one-on-one or pause the class and say something like, "Guys! We have an opportunity to Choose Again on our hands! Let's do it!"

2. Guide your student through these three steps:

   a. **Identify the negative thought:** Start by having the student identify their negative thought about math. In this case the thought was, "I suck at math. I'll never understand it."

   b. **Acknowledge and forgive the thought:** Encourage the student to pause and acknowledge the negative thought without judgment. What often happens is that students will acknowledge the thought and then get mad at themselves for thinking it in the first place, which is why the *no judgment* part is crucial. I tell students to not only forgive the thought, but to *thank* the thought for showing them what they *don't* want. For example, a student might say, "I don't WANT to feel like I suck at math; this thought is showing me that even though it's in my brain, it's not necessarily real but it IS showing me how I feel sometimes, like now."

   c. **Choose again!** Ask the student to "choose again" by selecting a "better" thought. That's usually the language I use, and I emphasize that the new thought needs to just FEEL a bit "better." The key here is that the new thought is still REAL and TRUE and feels AUTHENTIC. This could be a thought that acknowledges their current challenges but also includes a more optimistic perspective. So we're not going from "I suck at math. I'll never understand it" to "I am a math genius!!!" Instead, they might choose, "I may find math challenging at this second, but that doesn't mean I'll never understand it—there are lots of things I thought I would never understand that now I do!"

That's it! You're done! By using this Choose Again Action repeatedly and consistently, students learn to shift their mindset from a negative belief about their math abilities to a more positive and growth-oriented perspective, ultimately using those new thoughts to makeover their math stories!

 **Grade-Level Modifications**

- **Grades K–2:** Give students prompting, in the moment, to come up with new thoughts in case they don't come up with them on their own.

- **Grades 3–5:** Carry this Action out by having students simply say their old and new thoughts out loud, in the moment!

- **Grades 6–8:** Have students write out their old and new thoughts in their Math Therapy journals so that they can refer to them the next time they have a similar thought.

- **Grades 9–12:** Allow students to work with a classmate on reframing old thoughts into new ones. Sometimes an outside perspective can help students see their own thoughts with a new outlook!

 Makeover *Action 2: Fact Check!*

Why

Sometimes we repeat a story so often that we start to believe it, even if it's factually untrue! Remember, repetition can be our bff—but it can also be our worst enemy if used to embed false thoughts into our brains! By playing detective, we can help our students makeover their math stories in real time by helping them adjust the "facts"!

When

This is a quick Action you can jump into as needed!

How

1. Actively listen! Next time you hear a student say something like, "I've never been good at math," or "I literally always make mistakes," it's your time to STEP UP AND SHINE! This is your moment to pull out your magnifying glass and detective hat and fact check!

2. Play detective! Fact-checking doesn't involve screaming, "LIAR LIAR," to your student; it involves pausing and reflecting and saying something like, "Hmmm . . . is that actually true, though?" Jo Boaler (2019) has a strategy she uses called "Prove It to Me!" where she asks kids to be skeptics. She uses this strategy when solving math problems, to introduce kids to the idea of proofs. The idea is to have kids deeply interrogate the math they are doing to understand why it works in order to explain it to someone who might be a skeptic! This Action makes use of this same concept, but applies it to fact-checking our math stories to make sure we're not telling ourselves harmful lies. If a student tells you they ALWAYS make mistakes, then they need to prove it. Is there counterevidence that suggests that at some point they did NOT make a mistake while doing a math task? If so, they're spouting fake news and need to give their math story a makeover right then and there—voilà! If they just make a little tweak to their story, they can carry on with a new narrative in mind!

Grade-Level Modifications

- **Grades K–2:** Go gentle by being the one to offer alternative takes on students' math stories instead of asking students to help each other out. This will give you more control over empowering students in ways that are helpful instead of potentially harmful!

- **Grades 3–5:** Create a Justice League vibe in your classroom by letting students know that it is their job to help ALL students see the best in themselves! When a student makes a comment that reinforces their negative math story, call on the Justice League to help that student see their strengths instead of perceived weaknesses!

- **Grades 6–8:** Now that you're fostering a Math Therapy classroom, you can encourage students to actively help one another see their strengths in math class. Let students know that the next time a classmate expresses frustration or defeat, they have an open invitation to step in with counterevidence to help their pal see that they're actually doing a lot better than they might feel they are in the moment!

- **Grades 9–12:** It's hard for students to see the forest for the trees, so be the one to take charge and help to provide counterevidence in times when all they can see are their perceived failures. You can ask classmates to jump in with statements like, "Guys, am I wrong here or did Sasha not ask the BEST question last week?" Read the room and do what feels right!

 ## Makeover *Action 3: Switch It Up!*

Why

In this Action, students "switch spots" with one another so that they can hear their math thoughts reflected by another voice. When you hear your experience repeated in someone else's voice, you are more likely to be able to see it objectively. Role-playing allows students to examine and makeover their math narratives with a clearer perspective!

The exciting part of this is that as you continue to cultivate a Math Therapy classroom, students will not only be more comfortable jumping in to help a peer role-play their math stresses, but they will be EXCITED to do so! In fact, if you decided to do the Accountability Buddy Action from Step 3 (p. 156), you can add this to the list of things they can do with one another. This Action adds even more trust and collaboration to the Math Therapy culture you're cultivating—it's a win-win!

When

This is a quick Action you can have students jump into as needed!

How

1. Actively listen! Whenever a student expresses a thought that suggests that the negative aspects of their math story might be popping up, pause for a role-play! Don't do this in front of the whole class unless your classroom culture allows for it (this is your call!); instead, ask the student to buddy up with someone for a role-play! They can choose their accountability buddy, a friend, you, or let you choose someone for them.

2. Let the role-play begin! Don't worry—this isn't drama class, and we're not expecting Oscar-worthy performances. All that is required is for students to mirror each other's thoughts so that they can objectively react to what they're thinking. So for example, if Carli is saying things like, "I suck at math—there's no point in trying; I'll never get into the college I want, so what's the point?" the action is for Sabina, their classmate, to repeat a similar sentiment, as though they're the one who is feeling it. The directive is for Carli to objectively listen to Sabina as though the thoughts were her own and to pep talk her the way she would a bff. Carli might say something like, "Even if you don't get into your top choice school, that doesn't mean you won't get in anywhere," or "You don't suck at math; it's hard for all of us, but the more you try, the better you'll get, no matter what!" The act of coming up with that pep talk *seemingly for someone else*, combined with the action of articulating it out loud, will go a long way in allowing Carli to internalize it for her own benefit! It leaves students feeling gushy and good and not only helps them individually, but builds morale as a collective.

Grade-Level Modifications

- **Grades K–2:** Instead of role-play, try using the Choose Again strategy from the beginning of the toolkit to get students to reconsider their narratives!

- **Grades 3–5:** At this grade band, I would recommend sticking to being the one that holds up a mirror for your students instead of having them role-play with one another!

- **Grades 6–8:** Pairing students up for role-play can totally work here, but make sure you set parameters around what the goal of this Action is and what appropriate responses might be!

- **Grades 9–12:** Walk around the room while students are paired up so that you can get an idea of what math thoughts are being expressed by your students. Feel free to jump in if you hear something concerning or have the perfect pep talk in mind. Try to do this in a way that makes both partners in the pair feel valued, as though you're coming up with the pep talk *together.*

Makeover Activities

Use the following Activities to engage students in making over their math stories on a deeper level! Remember, you HAVE TO—HAVE TO—do Activity 1. Trust me. Like, I can't MAKE you do it, but just take my word for it and do it!

  Makeover *Activity 1: Mathography Makeover!*

Why

By helping students recognize, interpret, and makeover their math story, they empower themselves with a new narrative that opens them up to building a better, healthier relationship with math and to let go of the weight of the past.

Listen up, you already know ALL about this Activity. Why? Because YOU DID IT AT THE BEGINNING OF THIS BOOK (see p. 1)!

If you need a reminder: In a mathography, individuals may discuss how their attitudes toward math developed, memorable math-related experiences, influential teachers, challenges faced, and how they overcame obstacles in learning mathematics. The purpose is to provide a more personal and narrative perspective on a person's mathematical journey, emphasizing the human and emotional aspects of learning math rather than just focusing on achievements or grades.

Making over our mathographies makes use of **deconstruction** (Kropf & Tandy, 1998), a narrative therapy technique that involves breaking down and examining the dominant narrative or story that an individual holds about themselves. The goal is to challenge and redefine these stories, providing space for alternative and empowering narratives to emerge. It's a process of dismantling negative self-perceptions and reconstructing more positive and nuanced self-stories!

When

This Activity requires one or two lessons plus take-home time, depending on how in-depth you want to go with it and what grade band you're working with. I personally like to give students extra time to REALLY think about and reflect on their math stories because this is such a personal Activity, and students sometimes need personal space out of the classroom to get vulnerable!

How

1. Create a *Makeover* Mystery! Start by telling students that they're about to get a makeover . . . well, sort of! Introduce the assignment, which is to write out their math story, and ultimately, to give it a makeover. Part 1 is just writing the initial story; the makeover part is a secret, and you'll tell them about it when the time comes. For now, the goal is just to get their stories out on paper! Explain what a math story is and share your original math story as an example. Give them time to write or to take home the assignment if they want privacy, and then ask them to turn it in so you can make sure everyone has a mathography to work with for the next part . . . *Makeover* time!

2. Time for a glow-up! Now that everyone has a mathography to work with, it's makeover time! Explain the rules: Rewriting our stories doesn't mean that we change the facts about our lives but that we consider different ways of looking at those facts. This is a good time to share YOUR made-over mathography, just like I showed you mine earlier in this chapter on page 184. Go back there now and take a look. See what I did? None of the facts changed, but my perspective changed. My new story is a lot kinder and puts me in the seat of power when it comes to my relationship with math! This will take time, but I really recommend that you have them hand in their initial mathographies, then go through and highlight what they wrote, using the following prompts as guidelines that might help them start thinking of how they might rewrite their stories. This gives them explicit direction, and it will help you understand what each of them are going through and what they're bringing into your classroom. Here are some prompts you might use, depending on the content of the mathography you're marking up:

    a. **Ask students to break it down:** If Kya wrote something general, like, "I'm bad at math, it's too hard," ask why they think that and what they meant. Did this idea come from a person, event, or experience? Get specific so they can target the misconceptions.

    b. **Ask students to challenge themselves:** Is everything in their story factually true, or are there some statements that are more opinion than fact? If so, ask them to consider whether or not their opinion might be biased or not. Is there a truer, kinder version of this opinion that more accurately reflects the truth? In Kya's example, perhaps they elaborate by saying that they feel bad at math because they usually get bad grades and never understand new concepts. Are there times when Sascha got a "good" grade or understood a concept? How can they incorporate that fact into their made-over story?

    c. **Ask students to be nice:** Ask them how they would rewrite their story if they were rewriting it for a friend. Would they be more objective? In Sascha's case, would they really say, "My friend is bad at math," or would they say, "My friend struggles with math, but actually there have been times when they've gotten a pretty good grade on something and have surprised themselves by understanding a concept before anyone else does!" Ask them to reexamine their story as if they were looking at someone else's story; they are likely to be kinder to a friend than they might be to themselves.

These are just some of the prompts you might consider, but your line of questioning will depend on what each individual mathography says! Remember, the point is simply to have students see that their math stories aren't necessarily objective, accurate representations of their entire math experience and that they have the power to makeover their mathography so that it empowers and inspires them!

HOT TIP

Remember to respect your own boundaries! If something comes up in a student's mathography that feels like it's beyond the scope of what you're comfortable with, turn to the protocol that your school or district has with regard to the physical and emotional safety of students. Never hesitate to seek out the support you need while doing this work.

3. Say it louder! Now, here's the catch: Your students have been repeating their OLD math stories in their heads EVERY day (remember what I told you about how we subconsciously repeat 90% of our old thoughts every day), and we all know that we need repetition to ultimately change! Tell your students to take their new math stories and stick them somewhere at home where they will see them every day. Ask them to read them out loud EVERY SINGLE MORNING for the next 30 days in a row. Check in after 30 days to see how it's going—are they feeling different?!

 **Grade-Level Modifications**

- **Grades K–2:** Students at this grade band don't have a huge history to draw on, so instead provide them with a few reflection sentence starters with options to fill in the blank with! For example: I like doing math (Circle one: never/sometimes/always) or I like trying new things in math (Circle one: never/sometimes/always).

- **Grades 3–5:** At this grade band, it might be a little tough to have students dig into their limited history. Instead, consider having them jot down adjectives or create a short poem that describes their math story. Help them rewrite by finding synonyms that are kinder and more aligned with growth mindset!

- **Grades 6–8:** Use story maps to help students structure their math stories. You can find a ton of these online, and I have included some of the most popular formats here!

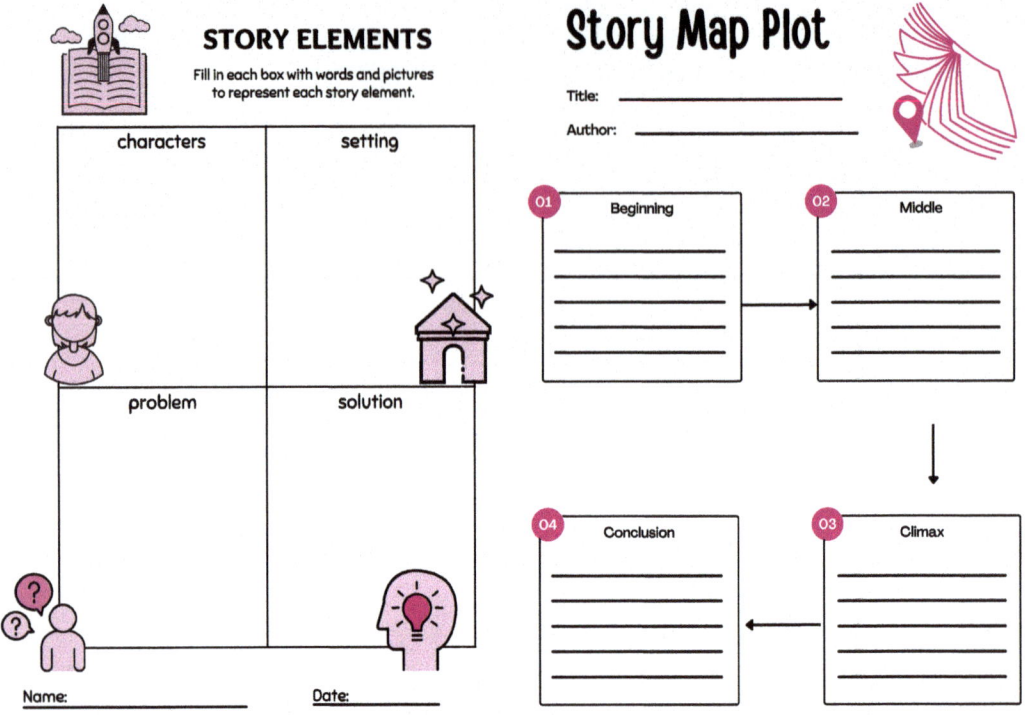

STORY ELEMENTS

Fill in each box with words and pictures to represent each story element.

characters	setting
problem	solution

Name: _____   Date: _____

Story Map Plot

Title: _____
Author: _____

01 Beginning

02 Middle

04 Conclusion

03 Climax

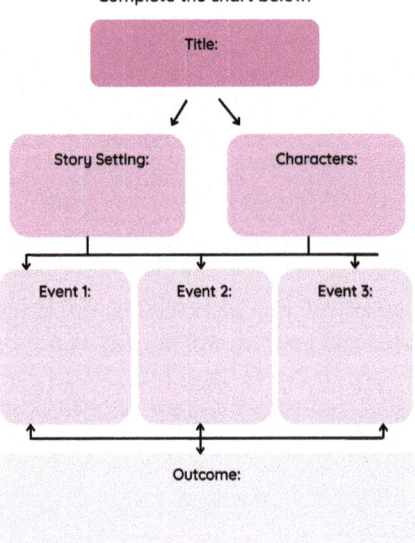

STORY MAP

Complete the chart below.

Title:

Story Setting: Characters:

Event 1:   Event 2:   Event 3:

Outcome:

- **Grades 9–12:** Allow students to choose between writing out their math story or using alternative media like video or audio to relay their math story in a creative way. (You might even have students interview one another, using the suggested prompts in #2 a, c as interview questions!)

 Makeover *Activity 2: My Math Sidekick*

Why

By helping students see their problems or behaviors as external to their core being or identity, those problems and behaviors are much easier to address and change.

This action is based on **externalization** (Hamkins, 2014), a narrative therapy technique that involves separating a person from a particular issue or problem. By externalizing the problem, we allow students to gain a new perspective and reduce the impact of the issue on their identity. This action helps students see that they are not *defined by* the problem! It's a lot easier to change a behavior you do than to change a core personality characteristic. For example, if a student considers themselves to be "a slow person" when it comes to math, then they need to fundamentally change something about themselves *as a person*, which is a daunting thing to think. BUT if a student, instead, shifts that thought so that they think of that attribute as a sidekick they're traveling with on their math journey (e.g., a math sloth that sometimes slows them down), now they don't need to change *themselves*; they instead need to work on changing their relationship to *their sidekick*. I know it seems whack, but it will make more sense when you look at the actual steps that follow—and it works, especially for younger kids!

When

Spend 20 minutes introducing the concept of a math sidekick and then ask students to refer to their sidekicks whenever negative thoughts about math threaten to take over!

How

1. Introduce it! Spend 10 to 15 minutes explaining the idea that our challenges with math aren't an inherent part of our core personalities, but more like sidekicks that are accompanying us on our math journey! You might want to discuss popular sidekicks throughout history and how even they sometimes challenged the main character! For example,

   - **Nicole Richie (Paris Hilton's sidekick in *The Simple Life*):** Nicole is a great bestie, and one of her greatest assets is that she can be down-to-earth when Paris goes astray. While Paris appreciates this about her, sometimes it takes the wind out of her sails—but the payoff is that she usually avoids doing things that could ultimately be detrimental!

- **Karen Smith (Regina George's sidekick in *Mean Girls*—the Amanda Seyfried character):** While she adds to The Plastics, her naivete and lack of awareness can interfere with the opportunity to address serious issues within the group.

- **Hermione Granger and Ron Weasley (Harry Potter's sidekicks):** Hermione's emphasis on rule-following and Ron's occasional lack of confidence sometimes slow down Harry's more impulsive and risk-taking tendencies, and their differing approaches create both conflicts in decision-making as well as creative solutions as to what steps to take next!

2. Your turn! Have students come up with their own math sidekicks by personifying their perceived math challenges. Students can illustrate their sidekick in their math journal or on a separate sheet of paper and answer the following prompts alongside their illustration. Here are the prompts they should follow in creating their sidekick, and you can do this in class or assign it for homework:

   a. Pick a negative thought you have about your relationship with math! *Example:* I'm slow at math. It takes me longer to understand concepts than others.

   b. Imagine that the negative thought you wrote down is like a character or sidekick. What would you name this character? *Example:* Sam the Sloth

   c. You are NOT your sidekick, but they ARE along for the ride. How do they influence your thoughts and feelings about math? *Example:* Sam the Sloth sometimes makes me feel like it takes a long time to answer questions and understand concepts when everyone in the class seems to be moving faster.

   d. What are some awesome things about your sidekick? *Example:* Sam is slow, but he's also really careful, so he's less likely to make mistakes.

   e. What strategies and approaches can you use to work with your sidekick better? *Example:* Instead of making Sam feel bad for being slow, I could thank him for taking his time and be patient with him. That would probably make both of us feel less stressed out when we're doing math together.

   f. If you had to write a one-sentence thank-you note to your sidekick, what would you say? *Example:* Thank you for helping me work at my own pace and not worry about how everyone else is doing. Slow and steady wins the race!

3. Bring it up often! Whenever you hear students reiterating their negative math perceptions, remind them of their sidekicks to bring this action back to life and remind them that our sidekicks don't define us—but they DO have the power to help us out along the way in surprising ways we may not have expected!

Grade-Level Modifications

- **Grades K–2:** To make this Activity relatable, have kids come up with a dream stuffed animal as their sidekick! They can still use the prompts listed to describe and expand on their chosen new stuffed animal friend!

- **Grades 3–5:** To simplify this Activity, have kids come up with a generic animal as their sidekick! They can still use the prompts listed in order to describe and expand on their chosen new animal friend!

- **Grades 6–8:** Make this an interdisciplinary Activity by giving students time and materials to get artsy with their sidekick! Have them turn their sidekicks into sculptures or posters that they can place on their desk or alongside them whenever they're doing math, that way they're constantly being supported by their trusty new pal!

- **Grade 9–12:** Allow students to use media, online images, or even AI to help them generate their sidekick!

⏳⏳ Makeover *Activity 3: My matho-GRAPH-y*

Why

By creating a visual representation of their math story, students are clearly able to see the correlation between past events and experiences and how those affected their narratives. Acknowledging and recognizing how their math stories came to be allows them to separate the past from the present, empowering them to rewrite their story moving forward!

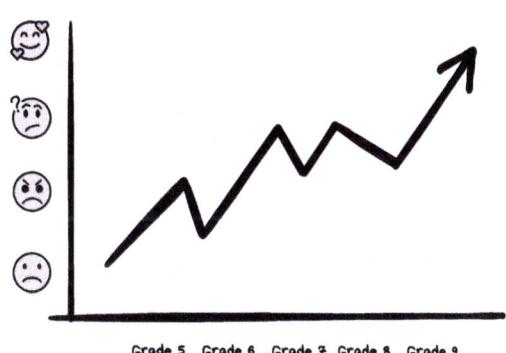

Grade 5   Grade 6   Grade 7   Grade 8   Grade 9

HOT TIP

If you love using tech, Desmos has collaborated with *Dear Math*'s Sarah Strong to create an online version of a similar Activity! Try it! You can find it here: https://bit.ly/4aTgMbJ.

One of the bonuses of this Activity is that it contains a mathematical component: the exploration of graphing! This opens up the opportunity to talk about correlation, cause and effect, how the x and y axis operate, and more, depending on what grade band you're working with. This Activity is super flexible and allows students to dig into their math pasts in a way that is creative, visual, and kind of math-y!

When

Dedicate a lesson to this Activity OR spend 20 minutes introducing the Activity and assign the actual matho-GRAPH-y for homework!

How

1. Introduce it! Spend 10 to 20 minutes explaining the concept of a matho-GRAPH-y. Tell students that they will be creating a graph to showcase their math story and go over some key features of a graph:

    a. The x-axis should represent time, and I would recommend using students' ages or grades instead of actual years (e.g., 6 yrs. old, 7 yrs. old, 8 yrs. old or Grade 5, Grade 6, Grade 7 vs. 1998, 1999, 2000) but up to you!

    b. The y-axis should represent feelings about math, and there are several ways to do this. I have seen some teachers put a scale of 1 to 10 on the y-axis (1 = I hate math, 10 = I love math), and I have seen others put various emojis on the y-axis. The cool part about this is either you can decide yourself, or you can make this a part of the Activity, which opens up a discussion about what certain emojis feel like (is it better to feel like 😣 or like 😥?!) and how those feelings translate to math. This is such an important discussion—we don't talk to our students about the complexity of emotions enough!

    c. Encourage students to create a minimum of one point for each number on the x-axis. So at each one of their ages, they need to find something on the y-axis that illustrates their math relationship at that age, place a point there (we call this a coordinate!), and label it. The label gives us all the info, so encourage students to write information on that label that describes why they chose that particular point—for example, something like, "This was the grade I didn't understand long division and then felt lost for the rest of the year!" If they need more space, they can write a full description elsewhere (I have included a template with a suggested format).

SOURCE: Emoji images from istock.com/bortonia

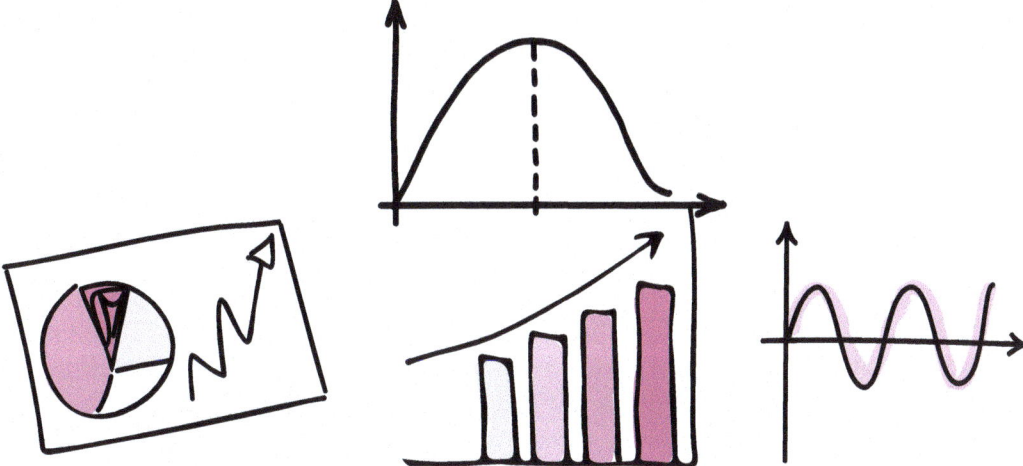

2. Follow up! Your students are ready to go. After they have turned in their matho-GRAPH-ies, you might consider following up one-on-one to ask questions about their math stories or to use one of the Actions in your *Makeover* Toolkit to help them revisit their math stories from a more objective perspective that allows them to detach from any of the negative storylines they've been carrying with them all this time! This Activity is great for helping students acknowledge their stories, and it's equally great for giving YOU insight into their past math traumas and more. What makes it even more powerful is taking it beyond acknowledgment *to transformation* by way of your feedback! Some feedback prompts you might consider include:

   a. Tell me more about this specific point on your matho-GRAPH-y!

   b. Why did you choose this emoji? What does it mean to you?

   c. You put a point next to a puke-face emoji when you were in Grade 5 and wrote that it was because you felt sick every time you wrote a math test. I'm so sorry—that must not have felt good! Do you still feel that way now? If not, what do you think changed? If so, let's talk about it and figure out why!

 **Grade-Level Modifications**

   • **Grades K–2:** Graphing is a far reach for this grade band, so instead have students use emojis to describe math memories or events, and don't worry about the timeline of when they felt what but instead on the memory of what happened and how it made them feel!

- **Grades 3–5:** Use emojis to create the *x* and *y* axis scales and involve students in the discussion of which emojis should go where! This is a great opportunity to open up a conversation about feelings around math, what they look like, and what they feel like!

- **Grades 6–8:** Start a discussion about cartesian planes here, so depending on what grade you're teaching, consider involving the negative *x* and *y* axis into this Activity—imagine how wild and fun that could be?!

- **Grade 9–12:** It's hard not to want to get really creative with graphing considering that functions are introduced at this grade band! Could we maybe even come up with . . . a math story . . . EQUATION? I'll just leave it at that . . . !

MAKEOVER TOOLKIT: SIMPLE SWAPS

Want to take your *Makeover* to the next level? Use these simple swaps to make subtle shifts that will have a huge impact on turning your students' natural negativity bias around! Negativity bias can manifest in various ways in a math class, including being laser-focused on the potential of failure, attributing success to luck instead of hard work, resistance to asking questions because it feels pointless, comparison to peers, and selective emphasis on what they don't understand vs. what they do. Be on the lookout for signs of negativity bias and then take these simple actions to give your students' negative self-talk a makeover in the moment!

SWAP THIS	FOR THIS
Telling kids that the point of solving a math problem is to get the right answer	Tell kids that you are just as interested in their process as you are in the final result.
Giving general feedback (e.g., "Great work!")	Give *specific* and *exact* feedback that highlights students' roles in achieving success. (e.g., "All of that practice I saw you doing yesterday paid off!").
Assuming that a disengaged student is just disinterested	Consider that they might be scared to try something new because they've already decided they'll fail and make it clear that they don't need to "get it" right away—that trying IS success!
Emphasizing that ease is the goal in understanding a math concept	Emphasize that hard work, practice, and perseverance are the *most important* skills a mathematician can have and that just because something is "hard" that doesn't mean you're not "good at it."

SWAP THIS	FOR THIS
Asking if anyone has any questions (and then dealing with the awkward silence of no one wanting to speak up . . .)	Tell students that sometimes you think things are clearer than they are and could use some help making sure you get your message across. Then, ask students if there is anything you might be able to clarify further for the class!
Moving on to the next thing the second one unit or concept is over	Take a moment to celebrate how far students have come! Say something like, "Guys, how cool is it that 2 weeks ago you didn't even know this concept existed and now you do? Look how much you're GROWING!"
Only highlighting where students went wrong on tests or assignments	Remember, they will focus on the negative, so make sure you also highlight where they went right and what they did well!
Using red ink to denote errors or mistakes	Use pink, purple, or orange—red is scary! We all know that, but also there's research to back it up (Walton, 2001)!

 ## *MAKEOVER* TOOLKIT: THEY SAY, YOU SAY!

When students get stuck in their outdated narratives, I have found that the following responses often go a long way in moving them along!

WHEN THEY SAY . . .	YOU SAY . . .
"I've never been good at math."	"What does being 'good' at math mean to you? Can you think of literally ONE time when you felt that way and what it felt like?" or "What if I told you that being 'good' at math doesn't mean what you think it means?"
"I have to work harder than everyone else to understand."	"So what? Michael Jordan got held back from the varsity high school basketball team for a whole year because his coach said he wasn't good enough yet. Did that ultimately make him a bad player . . . when it counted?!"
"Math doesn't come naturally to me."	"What does it mean for math to come naturally? Do you know that most people who love math have had to work really hard at understanding it? The world's most famous mathematicians spend YEARS trying to understand a SINGLE concept!"
"I only understand this concept because it's easy."	"Have you considered that other people don't understand this concept, so it obviously isn't that easy to everyone?"

⭐ *MAKEOVER* TOOLKIT: EXPANSION PACK

Having so much fun giving your students' stories a makeover that you want to keep going? Don't worry, I gotchu! The following are a list of more of my fave Activities. You can find templates and instructions for most of these Activities by heading to the online resources on the Math Therapy website, so let the *Makeover* continue!

1. **If I Were a Number:** So as I'm writing this book, there is a WILD TikTok trend happening right now where people are posting videos discussing the personalities that different numbers have. For example, one guy said that the number 4 seems chill and easygoing, but 7 just seems like it has an attitude problem, and 3 kind of feels awkward. Whatever, this is all made up, but the fascinating part is that this is *resonating* big-time with kids on the platform. People are going NUTS for this, and I think that it shows that there is a deeply emotional and personal component to math that ALL of us can relate to *if invited to the conversation*. So this Activity asks kids to pick a number—ANY number—that reminds them most *of themselves*. The catch is that they can't just pick a number without explanation—they need to point out what characteristics this number has that makes it so relatable and describe this number's special math superpower. So for example, Jason might be like, "I am totally the number 111 because 111 is a straight-shooter just like me; it's simple yet kind of cool, and if 111 were to do math, he would just say it like it is and not overcomplicate things—that's my style." Students will like this Activity because there is no right or wrong answer, and it allows them to insert positivity into their math story without making it feel like they're bragging about themselves!

2. **Five Things I Like About My Math Story:** Given our brains have a negativity bias, it is likely that our math stories will be riddled with more negative thoughts than positive. One way to balance that out with zero prep is to ask students to write down five things they like about their math story. Maybe they like that they worked hard to get where they are or that they had a teacher who believed in them or that they found that they liked trigonometry way more than they thought they would. Just jotting down five things that they are grateful for on their math journey so far will help them focus on the positive instead of solely on the negative!

3. **Make a Self-Talk Game Plan Board:** As a class, create a game plan for when students' negative self-talk spikes up. You can crowdsource feelings that your students experience or even take examples directly from their math stories or things you've heard them say. As a class, decide on an affirmation that everyone can say if they experience any of the common situations you've come up with together. For example,

when a student feels nervous, your class affirmation might be "Being nervous just means I care," or when a student feels discouraged, your class affirmation might be "I have something to gain from every experience, even if it doesn't turn out the way I had hoped!" Set up a game plan board and tack it on the wall so that students can flip to the affirmation they need in the moment!

STEP 4 TOOLKIT
TO GO
makeover

makeover
ACTIONS

makeover
ACTIVITIES

1. CHOOSE AGAIN!

Encourage students to reframe negative self-talk by choosing "better" thoughts!

1. MATHOGRAPHY MAKEOVER

Help students recognize, interpret and rewrite their math stories.

2. FACT CHECK!

Help students determine whether their negative thoughts are based in fact or opinion.

2. MY MATH SIDEKICK

Empower students to imagine their math challenges as superhero sidekicks!

3. SWITCH IT UP!

Have students "switch spots" with one another so that they can hear their thoughts reflected back to them.

3. MY MATH-O-GRAPH-Y

Allow students to graph their relationships with math!

makeover
EXPANSION PACK

IF I WERE A NUMBER ✹ 5 THINGS I LIKE ABOUT MY MATH STORY ✹ MY SELF-TALK GAME PLAN BOARD ✹

BEFORE WE MOVE ON

You have another step of Math Therapy under your belt—you are just CRUSHING IT! Time for a few to-go snacks and self-reflection treats before you move on to THE FINAL STEP!

Treat Yourself

Time for a treat to celebrate how far you've come! Since this chapter is all about rewriting your narrative with more kindness and compassion, this chapter's treat is about showing that same kindness and compassion—to someone else! Consider a small act of kindness that you can do for someone in your orbit. Make a donation, buy a coffee for someone behind you in the coffee bar line, or simply compliment a stranger. The smallest act of kindness can make the biggest difference and keeps the flow of compassion trickling through the collective; you are the first domino to fall, so be the change you wish to see!

ASK YOURSELF

1. Did this chapter change or challenge the way you think about math identity? Why or why not?

2. Before reading this chapter, did you think your math story was set in stone and couldn't be changed? If so, have you changed your mind?

3. Which of your students do you think will benefit most from the Actions and Activities suggested in this chapter? Which do you think won't benefit as much? Why?

4. Which Action or Activity are you most excited to try with your students?

5. What is something in this chapter that makes you go, "aHA, this is totally going to make a difference"? Why?

6. What is something in this chapter that makes you go, "That will NEVER work with my students!" Why?

7. What are some of the challenges you may face when trying to implement the suggestions in this chapter? How might you prepare for these challenges ahead of time?

☮♡π

CHAPTER 7

MATH THERAPY STEP 5

Measure Progress

In this chapter, we get to:

1. Learn the truth about progress and perfection

2. Get creative with how we assess our students

3. Discover the importance of celebrating every little win

4. Dig into our Step 5 *Measure* Toolkit and empower ourselves with impactful Actions and Activities to help students see how far they've come on their Math Therapy journey

I once had an interaction with a student that baffled me to the point of speech-lessness (something which, I'm sure you can imagine by this point, is VERY rare for me). Sydney was crying on the couch of my tutoring studio because she hadn't gotten into her top-choice college. I had been tutoring Sydney for 3 years at that point, and she had gone from, "I will never be able to do math and I hate it with all my guts" to applying to STEM programs across the country and even debating the possibility of becoming a math teacher one day. She even begged me to take a pic of her first A on a math test and feature it on my Instagram account. She had worked really, really hard. However, on this particular day, she was sobbing and back to blaming math for all of the problems in her life, the latest one being that her grade hadn't been high enough to secure her a spot at her top-choice college. Now, I was doing all of the things you probably would have done. I explained that I hadn't gotten into my top-choice university, and here I was, living my best life! I told her that I literally didn't know a SINGLE adult who was like, "Man, my life would be so much better if ONLY I had gone to _____ college." Finally, I tried getting philosophical with the whole "whatever-is-meant-to-be-will-be–everything-happens-for-a-reason" approach. Nothing worked. Finally, through tears, she said the words that would render me speechless:

> *"What's the point of all of the hard work I've put in if it didn't get me the result I wanted?!"*

This is actual footage of what my face looked like, btw:

I actually couldn't believe what I was hearing. Sydney thought that the only point of working hard was . . . getting what you wanted? It frustrated and saddened me. After a few moments of literal stunned silence, I finally said,

> Sydney, the point of all of your hard work over the past 3 years wasn't just to get you into one specific college for a few years of your life; the point is WAY bigger than that! Think of the way you think of yourself NOW vs. then! You're more confident, you're more courageous, you believe in your capacity to change—all of THOSE things were the point, and those are the things that matter the most because they will be with you for a LIFETIME!

I'll be honest, as rousing as I thought my speech was, it was met with a blank stare. And then more sobs. She didn't get it, she didn't believe it, and I didn't blame her. For her entire career as a student, she had been told that the only way to be successful was to get into a top-choice college, and the only way to do that was to get good grades, and therefore, the only point of working hard in math class was to get the good grades needed to get into a top-choice college. THAT was what success meant to her because that's what she had been taught.

Never fear: The story has a happy ending! Years later, Sydney and I met for coffee. She had graduated from college (Note: NOT her top-choice college, but did Syd even remember that? Nope!). She was going to teacher's college the next year and said that she literally could not have imagined things working out better. The past 4 years had been amazing, she had learned so much, and all of the hard work she had done in high school had paid off more than she had imagined: She had gone into college with incredible study skills that carried her through tough exams, the confidence to join clubs on campus and make new friends, and the resilience to keep at it when her morale was low. When I reminded her that she had once asked me what the point of hard work was if it didn't yield the exact result that it was directed toward, she paused and said, "Hmmm. You know, it sounds like something I might have said, but the truth is you just can't think like that because whatever will be will be—and everything happens for a reason!"

Actual footage of what my face looked like, yet again, all those years later!

STEP 5: *MEASURE!*

Guys, this is it! The final step! This is your chance to show those kids that *this is working!* What do I mean? Well, in Steps 1 through 4, students have worked through their complicated relationship with math, and hopefully, they're now feeling a tiny glimmer of hope: They CAN do it, they CAN feel better about math, and it starts NOW! In Step 5, it's time to show them that they *have* improved—not just in their hard math skills, but in their math-adjacent skills like risk-readiness, resourcefulness, and resilience. Even though this journey has been primarily about building a better relationship with math, your students' math skills will likely improve *as*

well because as I said a million years ago at the beginning of this book, that's usually the by-product of building a better relationship with the subject! How cool is that?!

In this step, we will explore Actions and Activities to do just that. Remember to always go back to the GOAL of Math Therapy:

Our collective goal when employing Math Therapy is *to empower students to build a better relationship with math, and in doing so, with themselves!*

Keep this top of mind as you explore this chapter. Keep your eye on the prize, and your students will do the same!

> **Even though this journey has been primarily about building a better relationship with math, your students' math skills will likely improve as well because that's usually the by-product of building a better relationship with the subject!**

MEASURE

The fifth step of Math Therapy is all about finding creative ways to show students that their Math Therapy journey is paying off. When we started Math Therapy together, the promise was that all students would build a *better* relationship with math, whatever that might look like for them. In this step, you will help them measure their progress in ways that illuminate the progress they have made on their math journey, no matter where it is that their own personal journey started!

Why It Works: The final step of Math Therapy works because most humans are fueled by results. When we see evidence that something is working, we are more likely to find the motivation to repeat that something, even when the going gets tough, because we have proof that it is getting us somewhere. This goes all the way back to our pal B. F. Skinner (1963)—remember him?!—and his operant conditioning theory. Basically, Skinner says that behaviors are strengthened or weakened based on the consequences they produce. Positive reinforcement is a KEY concept in operant conditioning, and the idea is that a rewarding stimulus after a behavior increases the likelihood of that behavior being repeated, which is precisely why we want to SHOW kids that the work they've been doing on their Math Therapy journey IS working!

> **When we see evidence that something is working, we are more likely to find the motivation to repeat that something, even when the going gets tough, because we have proof that it is getting us somewhere.**

How It Works: The Step 5 *Measure* Toolkit transforms research about the value of measuring progress into a set of practical strategies you can easily integrate as part of your current classroom practice. You will walk away from this chapter with Actions and Activities that will enable you to both measure students' progress for your own knowledge, but more importantly, to show students that they have progressed in building a better relationship with math! These strategies will tackle both the measurement of progress in math skills, as well as math-adjacent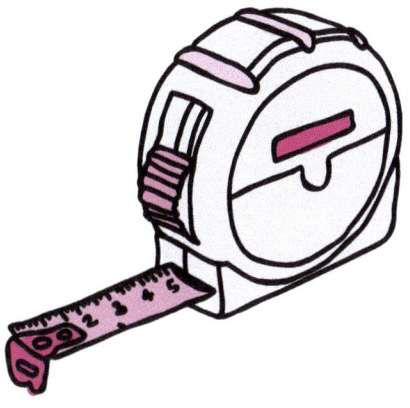

skills. So for example, we're going to show kids that even if they might not be able to fully add fractions today, at least they can find a common denominator—which maybe they couldn't do last week! And even if they still feel nervous in math class today, at least they're showing up—which maybe they weren't doing last week! Small and steady steps are our focus in Step 5. Remember, progress, not perfection!

Guys, this is THE LAST STEP. Are you freaking out WITH EXCITEMENT?!?! Now look, as educators, you are already experts at measuring progress, and you'll likely have noticed that in EVERY step opportunities exist to help kids celebrate their wins along the way, so you're already primed and ready for this section! In Step 5, we're going to take everything we've learned along the way and come up with concrete tools to help kids explicitly measure their progress in a way that is salient to them so that they can *hang on to those wins* despite that pesky negativity bias we talked about getting in the way! You can use ANYTHING you learn here at ANY time, just like with all the other steps. And of course, I wouldn't be me if I didn't remind you one last time that, just because we're at the final step, it doesn't mean we're at the finish line—because a finish line is just a goalpost that continues to move forward along with our journey as it progresses!

Now, before we get to the contents of our FINAL TOOLKIT, I want to share a few last thoughts about what the catchphrase "progress over perfection" actually means, why measuring progress is important, and how to C-E-L-E-B-R-A-T-E in ways that show students that when it comes to winning, size doesn't matter!

PROGRESS > PERFECTION

Our students have likely all heard the whole "progress over perfection" thing a bajillion times, but I wonder if they really understand the core essence of what it means? I say this because while those words have likely appeared on their social media feeds or on a classroom poster, a similar adage—"Actions speak louder than words"—has probably taught them that, while we tell them that progress is valued, in math class we only "officially" celebrate perfection! I recently did a poll on Instagram, and over 100 parents said that their child will get zero marks on a math test if, even though they've provided some correct steps to solve a problem, they haven't ultimately achieved the right answer. So in a sense, kids almost everywhere are still being shown that, if they don't get something PERFECT, the rest of their work is worth nothing.

Progress Is the New Perfect

I don't know about you guys, but every year I have a ton of students who have decided that "failing" means "getting less than an A." Last year, a student sent me a video of a bunch of girls in her class having a "crying party"—you read that right—after they had all gotten 80s on their tests instead of 90s. The weirdest part about this was that, like, this was a publicly known thing, and they had no issue taking photos and videos of themselves crying and posting them on Instagram with captions like, *Never getting into college at this rate* or

When you've studied till you're almost dead but still fail your math test. Like, good for them for being so open with their emotions, but come on, tell me that's not a little strange? There is a CULTURE around perfection that has developed in an insidious and detrimental way, and I can't even pretend to know where it came from exactly (I have a lot of unproven hypotheses, though!), but I can say that every teacher I've spoken with has agreed that, yes, students' standards of what qualifies as a success gradewise have definitely shifted in a way that's detrimental to their mental health.

Now, while crying over scoring an 85% on a math test may seem a little cray, I actually kind of understand the mentality here. The truth is that it's harder now to get into top colleges than it was, say, 20 years ago (Hoffower, 2018) for a variety of factors—an increase in the number of applicants being just one of them. In 1988, the acceptance rate for Columbia University was 65%; as of 2021, it was 4.1% (Columbia Undergraduate Admissions, n.d.). Likewise, the University of Michigan's acceptance rate dropped from 52% to 20% (PrepScholar, 2017) in the same time period. This has led to a lot of finger pointing around grade inflation and questionable grading practices. While perfect grades coupled with rejection from college has been a common occurrence in the United States for many years—and continues to be a big stumbling block—this is a new issue for most Canadians, illustrating that this is a problem that's only getting bigger. Last year in Ontario, not only did six kids from a Toronto high school graduate with 100% averages (Hurley, 2023), but the CBC reported that students with exceedingly high grades still weren't getting into their programs of choice (Crawley, 2023). One such student had the equivalent of a 4.5 GPA and was president of their high school student council and still didn't get admitted into business degree programs at the University of Toronto, Queen's, or McMaster (which are three of our "top-choice" universities over here in Canada). Many students have reported similar experiences, so it's kind of not THAT cray after all that they're shedding tears over an A, especially when they're being told that the be-all and end-all of high school math class is to get them into a top-choice college. Really, their quest for perfection is kind of an "us" problem if you think about it, which means it's also up to us to fix it!

How do we do that? The way I look at it is that it's crucial to help our students understand the freedom of not being able to ultimately control the outcome of every single situation. I know this might sound bleak, but it's actually very empowering! Focus on what you can control (trying your best, progressing little by little) instead of what you can't (getting into your top-choice college; even with a 100%, you might not achieve that outcome, and that's okay). This has a lot to do with what we talked about in Step 3 when we addressed internal vs. external motivation. All we can control in this life are *our actions*. We can't control other people's reactions, but we can place our focus on doing our best in every circumstance, giving ourselves grace, and enjoying the journey—because the destination isn't always going to be the one we had in our travel itinerary, and that's okay: We'll deal with it when we get there!

> **The destination isn't always going to be the one we had in our travel itinerary, and that's okay: We'll deal with it when we get there!**

HOT TIP

Watch your messaging! Make sure you consistently and repeatedly let your students know that the journey is just as important as the destination and make sure that your *actions* reinforce that message. By giving marks for progress and compliments for process, you can show your students that you really are practicing what you preach!

Practice What You Preach

Last year, I was invited to meet with a group of high school students to discuss their views on failure, growth mindset, and assessment. The principal told me that they were struggling with a student body that was highly anxious and so scared of failure that this translated into a lack of risk-taking in math class and intense panic at the thought of scoring anything less than perfect on a math assessment. So in I went, ready to get

everyone ALL PUMPED about the idea of learning from our mistakes and embracing failure and lalalalalala—and wow, was I in for a shock. One student just laid it on the line:

> We get told over and over again that progress is important and it's okay to fail, but how can we really feel that way if we need a specific grade to get into college, and to get that grade, we can't *actually* fail?

I mean . . . how do you argue with that exactly? I couldn't. They went on to explain that, even though their teachers theoretically believe in growth mindset, it felt like that philosophy didn't seem to apply to assessment. Like, growth mindset was all the rage—until it came down to the actual business of grading, at which point it seemed to go out the window. I dug deep into my inner Crystal Watson (2024), who is a huge advocate for the role of student voice, and suggested they set up a meeting with the head of their math department to discuss because they were making valid points and deserved to be heard. And so they did! The result was an adjustment of assessment practice and rubrics by the math department in order to honor the philosophy of growth mindset they were promoting and to allow students to explore math in ways that would allow them to take risks and think creatively without the fear of being penalized. Some of the adjustments included the elimination of one-shot assessments, the addition of opportunities for students to demonstrate knowledge in a variety of formats, and the inclusion of collaborative assessments to honor the value of collaboration to the process of mathematicizing.

While we don't all necessarily have the power to totally overhaul our assessment practices, there are some actions we can take to nudge our practices closer toward a growth mindset philosophy, so let's talk about some of the things you CAN do, and as always—start where you are, and do what you can—progress over perfection!

IN THE MOMENT

Consider your most frequently used assessment practice. How might it inadvertently disadvantage certain students? Are there ways to tweak this practice so that it offers all students the opportunity to demonstrate how much they've truly learned in your classroom?

Let's Talk About Assessment

My friend and fellow math educator Melissa Dean once said,

> *We can't preach a growth mindset and then assess from a deficit mindset.*

What a mic drop! You know, when we started our Math Therapy journey together, we talked about how students don't pay attention to what we SAY as much as they pay attention to what we DO. If we preach growth mindset and then assess from a deficit perspective, it makes us seem like we're full of sh*t. Circling back to Step 1, if we want to *Mythbust* the idea that

> *If we preach growth mindset and then assess from a deficit perspective, it makes us seem like we're full of sh*t.*

perfection is the whole point of math class, we need to change our assessment practices and the way we define success. Now, we've done a lot of that in Steps 1 and 3, but now I want to straight-up talk about assessment and how we can get creative with it within the red tape that may or may not define our boundaries. This conversation applies to both formative AND summative assessment because both are significant in showing students not just what we value, but what we want THEM to value as well!

Be Flexible

One of the best ways to allow students to truly feel as though they're making progress is to allow students to retake assessments. This shows them that progress really is more important than getting something right the first time. There are several ways to accomplish this. The first is to allow students to retake a test multiple times, ultimately getting rid of one-shot assessments. I know it seems like this will be time-consuming for you,  who has to mark each of these assessments, or that you'll have students getting 90s who want to retake the assessment over and over again until they finally get it perfect, but most teachers who do this report that this kind of abuse of a retake is actually fairly rare. That being said, if this is something you're worried about, you can set parameters—for example, let students know that they can only retake assessments if they score less than a specific mark on their first attempt or set a limit on how many retakes are allowed. I would suggest actually coming up with these parameters *as a class*; that way students feel involved in the process, and their voices feel respected and valued! In fact, you might even consider adding this to the classroom agreement you may have made in Step 2—what a full-circle moment!

Remove Barriers

Recent research has found that in the United States, millions of kids that actually need Individualized Education Programs (IEPs) don't ever get them (The Understood Team, 2017), due to factors including eligibility, cost, and

extensive wait times. This means that there are kids in our classrooms who might have the opportunity to develop an entirely different relationship with math if simply given access to accommodations like extra time on tests, formula sheets, or the opportunity to demonstrate their knowledge in creative ways. So what does this mean for us as teachers?

If you can remove barriers, do it. Period. If a student doesn't have an IEP but you're finding that they consistently run out of time on tests, if you can, give them extra time. If a student knows what they're doing but is struggling to remember formulas, give them access to a formula sheet. If a student gets anxious and distracted by the sound of other students' papers rustling, allow them to use noise-canceling headphones or to take their assessment in a private space. I'll even go one step further and suggest that you just make accommodations *for all students* if you can (within reason, of course!) because it levels the playing field AND it shows your students that you're an ally in their math journey AND that you don't want things that have nothing to do with what you're assessing to get in the way!

HOT TIP

If you're interested in removing barriers for all students in your classroom, here are some ideas for you to consider:

- Flexible instructional methods: Use a variety of instructional methods, such as visual aids, hands-on activities, and real-life examples, to reach all students!

- Extra time: If you can, allow students who need additional time to complete assignments and assessments—it can make a big difference!

- Visual aids and technology: Provide access to assistive technology—such as calculators, speech-to-text software, or screen readers—for those who need them!

- Flexible seating: Allow for flexible seating arrangements to accommodate different sensory needs and learning preferences.

- Graphic organizers: Provide graphic organizers to help students organize information and solve problems step by step.

Keep Your Eye on the Prize

One of the hottest debates out there right now is the one surrounding calculator usage in math class. Half of my Twitter feed is full of folks screaming that the reason kids don't know their times tables is because they're all using calculators, and the other half is full of responses along the lines of "Kids have calculators on their phones, deal with it." Regardless of where you are on the screaming-on-Twitter-about-calculator-usage spectrum, the point is that we always need to make sure that our assessments are truly showing our students what we VALUE and that we're being honest in terms of what we're actually assessing.

If you're trying to assess whether or not kids understand how to use the quadratic formula, does banning the use of calculators truly serve your purpose, or does it create an unnecessary obstacle between your students' understanding of quadratics and their ability to accurately demonstrate that knowledge? Now, you might be like, "No, Vanessa, banning the use of calculators DOES serve my purpose, so stop judging me," and if so, I'm totally open to that, promise! All I'm saying is that over the years, I have had so many students who have felt totally deflated and discouraged because, even though they worked their butts off to master a concept, they never felt the results of that hard work because something that may have not been enTIRIEly relevant got in the way.

If a student has always had trouble with say, mental math, yet they totally understand how to solve a quadratic, it doesn't make sense to keep shining a spotlight on that deficit. Doing so keeps kids feeling like there is no point in trying and reinforces the myth (which we busted in Step 1!) that there are only certain ways to be good at math, and if they're not adhered to, then forget it. In the case of calculator usage, many of the teachers I have worked with have divided their assessments into two sections: one in which calculators aren't allowed and one in which they are. That way, any assessment of arithmetic strategies can be done in conjunction with the assessment of broader concepts that are also being evaluated. This is just one example, but it's one that comes up A LOT! Pay attention to what you're trying to assess and whether that's honestly coming across in your assessments. We ARE on our students' side. Even though we never mean to, sometimes if they lose marks for things they aren't aware they're being evaluated on, they feel "tricked"—and that erodes trust. You've got this! Do what you can, be honest with your students, and most of all, be honest with yourself!

HOT TIP

In his book, *Atomic Habits*, James Clear (2018) talks about how goals are actually not as important as the *systems* we have in place to make those goals happen. For example, if your goal is to work out every day, the system you set up (laying out your gym clothes the night before, setting your alarm at an hour that allows you to work out before you start your day, etc.) is actually the most important part because without a rock-solid system your goals are just pipe dreams. I actually took his advice and am proud to say that I have now worked out 5 DAYS IN A ROW after working out ZERO times in 3 months—it works! Help your students come up with systems to make their goals a reality and make that (the journey) the focus instead of the goal (the destination)!

A SNAPSHOT OF PROGRESS

This is the fun part! This is where you get to show your students that it was WORTH IT! That they have progressed! Their brains have grown! They have a better relationship with math TODAY than they did YESTERDAY, and that was THE WHOLE GOAL! But what does that actually look like?

At its core, measuring progress is about finding quantifiable ways to show students that they have improved at a skill over a specific period. The process involves evaluating changes, improvements, or achievements relative to established goals, benchmarks, or expectations and making sure that evaluation is transparent! What follow are the components of measuring progress to keep in mind as we get closer to our toolkit!

Establish Specific and Quantifiable Goals

Our overarching goal is to help students build better relationships with math, and that will look different for every student. We've talked a lot about what success might look like for each individual student, so now is the time to reinforce the idea that every win, no matter how big or small, *is still a win*. A goal of getting an A on a math test is just as honorable as a goal of not crying during an assessment. You do you, and as always, START WHERE YOU ARE! Now, once you've done that, it's important to make that goal specific, measurable, achievable, relevant, and time-bound. You might remember from Chapter 5 that goals formatted this way are known as SMART goals (Doran, 1981) and are often more effective than goals that don't follow a similar format because when it comes to SMART goals, there's a clear plan in place. Check out the following table for some examples of how you can transform a regular ol' blah goal into a fun, exciting SMART goal!

GENERAL GOAL	SMART GOAL
Feel less anxious in math class.	Do box breathing once a day for the next 3 weeks and use box breathing any time I feel stressed in math class (head to Chapter 4 for a refresher on this!).
Get a good mark in math.	Increase math quiz scores by 10% within the next 6 weeks so that I feel more confident and prepared for the final exam.
Feel better about math.	Celebrate all the lessons I learn and progress I make over the next 3 weeks by writing each one down in my Math Therapy journal.
Become a math expert.	Raise my class average by 10% in the next 2 months by getting help from a peer tutor twice a week.
Do my math homework.	Attempt every question assigned for homework for the next 3 weeks.

Conduct a Baseline Assessment

How can we measure how far we've come when we don't know where we started?! Make sure that students know where they're starting so that they can not only set realistic goals but also so that they can actually SEE their progress when it happens! Not crying on a math test might seem like a silly ambition until you're like, "OMG . . . I have cried on EVERY SINGLE MATH TEST THIS PAST YEAR. I guess that actually WOULD be a huge achievement if I managed to NOT CRY after all!"

Check In Regularly

Measuring progress is most effective when done at regular intervals, which is why your toolkit is full of Actions to help your students measure their progress frequently and consistently, without much prep from you!

Our role is to help students see in themselves what they cannot, and while they may not believe in themselves, they won't be able to argue about their progress if you have data to prove it.

Collect and Compare Data

We often can't see what's right in front of us, and our students are no different. Our role is to help students see in themselves what they cannot, and while they may not believe in themselves, they won't be able to argue about their progress if you have data to prove it. This can involve quantitative data, such as test scores or completion rates, as well as qualitative data, including feedback and observations!

Communicate, Celebrate, and Continue

Communicating progress to our students is the best part about this whole step because it is the WAY we communicate progress that has the potential to create so much joy! Recognizing and celebrating achievements and milestones along the Math Therapy journey is important for motivation and morale, which is why we're going to talk about what celebration actually looks like in a second. Remember that Math Therapy is an ever-evolving process, so encouraging students to use insights gained from progress to refine goals and strategies not only reinforces that there is something to be gained at every point in the journey, but that ultimately, success is not a noun to be *achieved* but a VERB that has us consistently *in action*!

CEL-E-BRATE GOOD TIMES . . . COME ON!

You better have that song in your head now—because I do!

I have a thing for confetti cannons. I got in the habit of buying them in bulk from Party City because my band has ended EVERY set, for the past 400+ shows, with a song called "This Is Yours to Make" that is ALL about going after your dreams, and I always pop a confetti cannon when the final chorus hits. I'm telling you, if you haven't experienced the pure joy of a crowd erupting in literal GLEE under a shower of confetti, you're missing out. There's just NOTHING like it. So one day last fall, Shira showed up for her tutoring session. She was in Grade 10 at the time, and I had been tutoring her since Grade 8, and she had been barely passing math and experiencing panic attacks on every single assessment for almost 2 years. We had slowly been working on not just her math skills but on mindfulness strategies and thoughtwork to help heal her past math traumas, and we were on a path toward building a better relationship math—it had been a rocky road. Well, something amazing happened. Shira came in, sat down, and told me she had an announcement to make, but that it wasn't a big deal. She had gotten a B on her latest math test, and the weirdest part (according to her) was that she had felt "totally chill" and had tapped into her breathing exercises twice during the test when she felt panic rising, and they had worked.

NOT A BIG DEAL? Like, this was the BIGGEST deal.

I excused myself for a second, went to my closet where I keep all of my band's supplies, and found—you guessed it—a confetti cannon. Now look,

I only have confetti cannons in ONE size: the size that I use to shower a crowd of 500+ people in confetti. I took that cannon, headed back to where Shira was sitting, and screamed, "OMG, CONGRATULATIONS THIS IS A HUGE MILESTONE AHHH," and POPPED that baby right in the middle of my tutoring studio! I WISH you could have seen the faces of all of the other tutors and students—and most of all—I wish you could have seen Shira (see Figure 7.1). It. Was. Insane.

FIGURE 7.1 CELEBRATING!

• Yet another blurry incident of me celebrating with confetti in our tutoring studio! The confetti cannon was so successful with Shira, I decided to try it again with Kaleda, who was in Grade 11 math at the time! I took these screenshots from a video I took from a tripod of this all going down. I hope it helps you feel like you were REALLY there.

Did everyone think I was actually nuts? Yes. Did it make the BIGGEST mess, like, ever? Also yes. Was it worth it? DEFINITELY YES.

Celebrating is one of the—if not THE—most important aspects of Step 5. So often in life—and in the classroom—we hit a milestone, and instead of celebrating, we just move on to the next thing. As adults, we are all too familiar with this! We really do live in a world that urges us to want more and more, to no satiable end. I'm not going to get all existential here, but, like, capitalism, am I right?! If we constantly want, then we can constantly be sold to. It's a whole thing. But this trickles down into all aspects of our lives and leaves us feeling like we're always at a deficit and never satisfied. I saw this quote somewhere, and I don't know who said it, but it stuck with me:

"Once upon a time, you wanted what you have now."

So. True. Once upon a time, we thought, Well, when I have _____ I'll be good, but then inevitably, we get it (or we don't and we move on)—but either way, the goalpost moves. And that's okay. That's growth. But if we don't truly bask in the celebration of our accomplishments and milestones, we will always be left wanting from lack—instead of wanting from a place of abundance. We will always be left *needing* more to feed an emptiness—instead of *desiring* more because we are hungry for the adventure. This applies to everyone, including our students, which is why it is so important that we shoot metaphorical (or real, up to you!) confetti cannons every time they hit a milestone so that they can stare in awe and wonder at the flurry of magic that illuminates what THEY have accomplished.

> *If we don't truly bask in the celebration of our accomplishments and milestones, we will always be left wanting from lack–instead of wanting from a place of abundance.*

If you're not into my whole abundance/deficit spiel, I'll let you in on the science of celebration. Remember earlier in the chapter we talked about operant conditioning and how individuals are motivated by evidence that their efforts are yielding results? Here's how it works:

When you do something and get positive feedback, you're likely to do it again. Why? Because your brain's dopamine system, which is linked to pleasure, reinforces the behavior (Hamid et al., 2015). Since dopamine is a neurotransmitter associated with reward, positive outcomes release dopamine, strengthening neural pathways, and creating a motivation loop. Now, remember that whole footsteps-in-the-snow thing we talked about in Chapter 3 (not to be confused with the MURDERY footsteps in the snow discussed in Chapter 6)? Let me remind you: In Chapter 3, we learned that our neural pathways are kind of like footsteps in the snow! The more we repeat the same path—or the same behaviors—the easier it is to walk that same path because the snow becomes nice and hard and packed over time, making it easier for our shoes to slip right in! But now that we're bringing the idea of positive feedback into the mix, not only is that snowy path EASIER to walk, but it's also sprinkled with little treats along the way (see Figure 7.2). Now

FIGURE 7.2 MATH TEACHER IRENE BLAZEK

• Irene has been a math educator in Iowa for 16 years and loves to pull out her pom-poms (from her cheerleading days!) to celebrate student success in her classroom!

SOURCE: Irene Blazek

when you walk the path, not only is it more familiar, but every time you reach the end, someone hands you a hot cocoa to say "congrats." That hot cocoa reward is kind of like that dopamine hit! Ultimately, the brain connects this repeated behavior with positive results, making it easier to repeat. This is why having proof that your efforts are actually paying off boosts motivation and makes you more likely to keep on going when the going gets tough (let's face it—if you're Canadian like me, it's hard to say no to hot cocoa)!

HOT TIP

Remember that while the use of rewards can be a useful tool, to keep things exciting it's good to have different kinds of rewards and not just the same ones all the time. Desensitization to dopamine can occur when the brain, exposed to high levels of the neurotransmitter, adapts by reducing receptor sensitivity or altering reward circuitry. Simply put, when we consistently get rewarded, our brains tend to get used to it and not feel as excited anymore. Also, make sure the rewards aren't always extrinsic (like stickers and pizza parties), but also intrinsic (like helping students feel a sense of pride or satisfaction in their work)!

Whether you call it dopamine, abundance, or hot cocoa, we can likely all agree on one thing: Celebration is SO IMPORTANT. And fun. And delicious. Wow, I'm high on dopamine just WRITING this, to be honest, and I'll actually be HIGHER once my editor gets back to me and tells me that I have written an AMAZING CHAPTER (see, I need validation and celebration to motivate me to keep going too!).

PUT *MEASURING* INTO PRACTICE

Are we ready to show our students that they have made progress on their journey toward building a better relationship with math OR WHAT?! This toolkit is SO MUCH FUN because it's full of Actions and Activities that will give your students that dopamine hit that will propel them to keep on going.

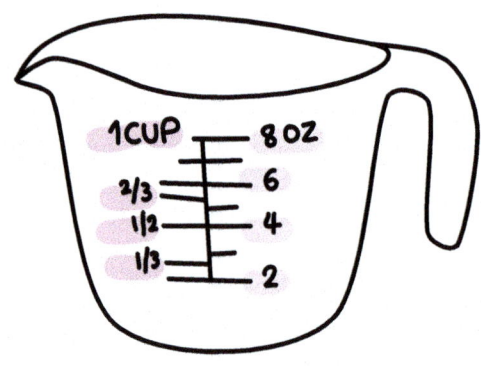

Before you finally get into this toolkit, let's take a moment to celebrate YOUR journey. You've done it. You've read right up until this point, and you are *still here*. Your journey isn't over, but you have achieved so many milestones along the way that I want you to soak it all in and feel proud. I am popping a confetti cannon for you RIGHT now, in my brain, as I type these words.

YOUR *MEASURE* TOOLKIT

Here I am one final time, reminding you that each toolkit contains *Actions* and *Activities*, quick swap-outs, and an expansion pack you can use to dive deeper into this step. You've heard it all before, so let's get right to the good stuff!

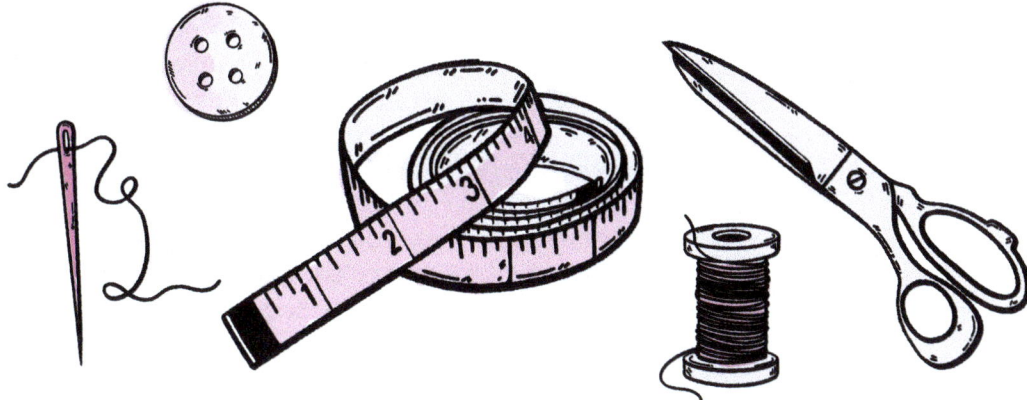

Measure Actions

Use the following Actions as a part of your regular classroom practice to help students both measure and celebrate how far they've come on their Math Therapy journey!

⌛ Measure *Action 1: Break Things Into Dopamine Bites!*

Why

Create multiple opportunities for students to both measure and celebrate the SMALL wins on their journey to understanding a BIGGER concept so that the resulting dopamine hits can continue to motivate them along the way!

Remember, dopamine increases motivation, so if you can break bigger tasks or math concepts into smaller chunks and then celebrate each of the resulting successes, your students will get the fuel they need to keep persevering! Also, as a bonus, breaking larger tasks or concepts into smaller bite-sized chunks is less intimidating and more manageable—it's a strategy I personally use when tackling a big project (like writing this book), and it works!

When

Do this any time you're introducing a new math concept or task!

How

1. Break it up! For example, if a big test is coming up in the next 2 weeks, create opportunities for students to take mini-quizzes or pass through knowledge checkpoints along the way. If you're teaching a unit on adding fractions, you might break it down into understanding what multiples are, finding common denominators, adding numerators, and reducing the final fraction—all of these chunks provide opportunities for celebrating mastery along the way!

2. Give them that dopamine hit! As educators, we're often used to breaking down our lessons and tasks, but not necessarily in making a big show of celebrating mastery along the way, and that's what I want this Action to add to your practice! Create opportunities for students to both measure and celebrate EACH progress chunk. So if you're ultimately teaching kids how to add fractions, consider how you might celebrate the student who totally mastered how to find a common denominator BEFORE you move on to talking about the step that follows. A quiz? A knowledge checkpoint? An "I'm so proud of you, this is a HUGE step" statement? A sticker for each step along the way to understanding a bigger concept? Often, we're sort of like, "Cool, you guys get that? Okay, next step," but consider taking it a step further by helping kids celebrate every step along the way to get that dopamine flowing!

Grade-Level Modifications

- **Grade K–2:** Use manipulatives and visuals to break concepts like number sense and number recognition into bite-sized pieces!

- **Grades 3–5:** Working on arithmetic? Break a concept like column addition into chunks, which might include figuring out in which order to stack the numbers being added, lining up digits, labeling columns (ones, tens, hundreds, etc.), and getting the final answer. Spend time on each part of the process and help students see that the achievement of EACH chunk is a success worth celebrating!

- **Grades 6–8:** This Action can be useful when working on solving any type of word problem. Break the process into chunks like highlighting key information, creating "let" statements, translating the problem into "math," and then ultimately solving it. Remember to celebrate EACH step of the process so that students can get that dopamine flowing!

- **Grades 9–12:** I love using this Action when I'm teaching students to factor (don't @ me—factoring is my FAVORITE unit, HANDS DOWN). Break factoring down into chunks like finding the least common factor, noting which factoring method would be best for each polynomial, the actual factoring part, and simplifying final solutions!

 Measure *Action 2: Create a Compliment Response System!*

Why

When we truly accept a compliment, we make space for the validation of our efforts and achievements. This nurtures a resilient and positive attitude toward personal development, which is exactly what Math Therapy is all about!

Did anyone ever teach you how to respond to a compliment? It seems like such a little thing, but if you're anything like most people I know, your response to a compliment is to brush it off or to make an excuse for why you DON'T DESERVE IT! How whack is that?! When we do that, it takes away from the potential to celebrate that moment and acknowledge our role in our own growth. By showing our students (and quite frankly, ourselves) how to respond to a compliment, we allow them to truly feel the magnitude of their progress!

When

Do this any time you compliment a peer or student and notice that they're brushing it off. Jump in—this is your time to help them celebrate!

How

1. Come up with response principles! We want our students to learn how to respond to compliments in a way that acknowledges their role in earning that compliment. The following are some prompts that might inform the compliment-response principles you set for your class:

- **Express Gratitude:** Simply say, "Thank you." This acknowledges the compliment and shows appreciation for the kind words—for example, "Thanks! I really appreciate your kind words about my progress!"

- **Acknowledge the Effort:** If the compliment is related to your effort or hard work, acknowledge it. This reinforces the idea that your accomplishments are a result of YOUR hard work! For example, "I've been working really hard on my math skills, so it's really nice to hear that it's paying off!"

- **Reflect on Growth:** Try to connect the compliment to your growth and the lessons you learned. This emphasizes the journey and your continuous development. For example, "I've learned some cool math tricks along the way, and it's fun seeing how much I've grown, especially given where I started."

- **Own Your Success:** Take ownership of your achievements without downplaying them! You rock! You're amazing! You are the star of your own show! For example, "I guess it's true that I practiced a lot, and now math feels easier for me, so thank you for saying something!"

2. Create opportunities for new responses! The unspoken part of this Action is that you have to start doubling down on your compliments, but I know you're probably already doing that, because you're amazing! Now, if a student responds to a compliment you give them by brushing it off, this is your chance to gently point it out and give them the option to choose a different response! Provide them with options and ask them which feels right. This is sort of a twist on the Choose Again method we learned back in Step 4, so you can build on that if you liked it!

Grade-Level Modifications

- **Grades K–2:** Create compliment stickers that kids can give to friends in the class to stick on their water bottles, notebooks, or whatever surface they choose! (Um—you might want to put some limitations on which surfaces they can sticker!)

- **Grades 3–5:** Create a classroom poster that illustrates your compliment-response system principles so that students can easily refer to it when giving or receiving a compliment!

- **Grades 6–8:** Have students practice giving and receiving compliments with a partner to get them used to what language to use with one another. You might create a sheet with potential compliments and responses and have students match them up in pairs. Students can then explain their choices to the group, generating an impactful discussion around compliments!

- **Grades 9–12:** Consider making a compliment jar for your classroom with potential responses written on little sheets of paper. If a student is stuck on how to respond to a compliment, they have the option of selecting a response from the jar!

⏳ Measure *Action 3: Keep Track of Unique Student Benchmarks*

Why

When students actually see tangible evidence of their own unique progress, they are more likely to internalize and be motivated by it!

As we all know, students—like us—aren't so great at measuring their OWN progress because they lose sight of where they started as that mental goalpost continues to inch forward! I have found myself saying things to friends like, "Can we take a moment to acknowledge that literally a year ago you were all, 'I will never find love, everyone hates me,' and now you're standing in your own abundance, feeling confident even though you're still single? That is a HUGE DEAL!" The response I usually get is along the lines of, "Huh. Wait, I actually hadn't noticed or remembered that . . . thank you for pointing it out!" We need to do the same for our students, and we need to acknowledge *their own unique journey and the progress they are making along the way!* When we do, it makes a difference!

When

Try to do this at regular intervals (e.g., schedule a weekly or monthly check-in with each student) or simply jump in with a benchmark reminder whenever a student feels like they're not making progress!

How

1. Conduct pre-interviews! These don't have to be formal, but the vibe here is to get an idea of where each student is on their math journey *now* so that you can remind them of how far they've come *later*. This might involve some of the Actions from previous steps—such as asking them to define success, pinpointing what math traumas they're contending with, or flipping through the reflection prompts you've given them. Note a few things for each student as a benchmark for measuring progress later on. For example, you might write something like, "Matt can't quite grasp the concept of multiplication yet, he's scared to ask questions, and he feels like he can't get better at math." That way, when Matt ONE DAY raises his hand to ask a question—BAM—you can step in and celebrate that milestone since you have kept track of that precious benchmark!

2. C-E-L-E-B-R-A-T-E! Depending on what kind of time you're working with, one option is to schedule one-on-one check-ins with your students and use those to help them measure their progress. Remind them of the benchmarks that you recorded and have them reflect on their progress! If this seems too time-intensive for where you're at right now, don't worry! There are opportunities every day to remind students of where they started and where they are now!

 **Grade-Level Modifications**

- **Grades K–2:** Create visual representations of growth so that students can visually see their progress. This might involve collecting stickers, coloring in pieces of a pie chart until the pie is complete, or anything that allows students to see their growth in a visual format.

- **Grades 3–5:** Help students see their progress by using concrete examples to illustrate growth. For example, if a student has traditionally struggled with confidence, during your one-on-one you might say something like, "Remember last month when you told me that you felt like you couldn't do math? I've noticed that lately you've been actually helping your friends with THEIR math work, and I am so proud of you for working hard to feel better about math and for using that energy to help your friends! Keep on going!"

- **Grades 6–8:** Consider solidifying your compliments and celebrations on sticky notes that students can keep in their math notebooks or Math Therapy journals. Think of this as a "collecting stickers" vibe, but instead, your students are "collecting celebratory moments!"

- **Grades 9–12:** Celebrate student progress with meaningful one-on-one conversations or heartfelt handwritten notes that show students that you authentically feel that they have progressed and grown!

HOT TIP

Along with tracking student benchmarks with regard to math skills, keep notes about your students' math attitudes and how they typically respond to challenge. This allows you to help them measure invisible progress—for example, the way they react to a tough question or impending deadline. Help them measure their progress by inviting them to revisit a concept or particular question they struggled with in the past. Prompt them to reflect on how they felt while doing the question. Confident? Scared? Stressed but less stressed than they felt the last time? Remember, an improvement in math attitudes IS growth, even if it's less visible than a grade on a piece of paper!

Measure Activities

Use the following Activities to help students measure and celebrate their progress on a deeper level!

Measure *Activity 1: Open Middle Math!*

Why

As important as it is for kids to see their *relationship* with math improving, it's just as exciting for them to feel like their *hard math skills* are improving! When we give students the autonomy to solve problems their own way and place emphasis on the process, they are more likely to measure success in their hard math skills in a way that doesn't only celebrate obtaining the "right answer."

I love Robert Kaplinsky's (2020) Open Middle Math because it's ALL about doing math in a way that celebrates the process, and in doing so provides tangible ways for students to measure their progress and feel the results of that progress. Open Middle Math involves math problems with a clear start and finish; however, the middle part is left open for multiple solution pathways, encouraging students to think critically, consider various strategies, and deepen their understanding of math!

When

Tweak the math problems you already assign so that they adhere to an Open Middle format or choose from hundreds of Open Middle problems that have already been created for you! Some teachers like to start each lesson with an Open Middle problem, and others like to make sure that each assessment includes at least one Open Middle problem as a core part of the assessment.

How

1. Dig in! You can find hundreds of free resources and templates online at https://www.openmiddle.com. There are Open Middle problems for almost every single K–12 math concept, and as you start to familiarize yourself with the format, you can create your own to suit whatever skill it is that you're teaching! I have provided one of my favorites, Close to 1000—check it out!

Exercise: Close to 1000

*Adapted from Robert Kaplinsky's Open Middle Math program

Directions: Using the digits 1 to 9 exactly one time each, place a digit in each box to make the sum as close to 1000 as possible.

Some ideas to get you started: (you don't have to use these!)

☐☐☐ + ☐☐☐ + ☐☐☐

First attempt: Points: ____/2 attempt ____/2 explanation

What did you learn from this attempt? How will your strategy change on your next attempt?

Second attempt: Points: ____/2 attempt ____/2 explanation

What did you learn from this attempt? How will your strategy change on your next attempt?

2. Measure and celebrate! Open Middle gives you a tangible inroad into measuring your students' progress on specific math skills as WELL as math-adjacent skills. Notice that in the example provided, there is space for math work, as well as for reflection about process and growth. Remember how important those little dopamine hits are and make sure you celebrate as many components of progress as you can. As much as kids will value their growth on their Math Therapy journey, they will want to see that their MATH skills have also improved too—even just a little. This gives you the opportunity to watch them do actual math and to note marked improvement, even if they don't ultimately get the correct answer!

Grade-Level Modifications

- **Grades K–2:** Use an Open Middle strategy with any math concept you're working on and allow students to work with manipulatives, visuals, and any tool they choose to get to their answer!

- **Grades 3–5:** Try starting with the following Open Middle task:

Directions: Using the digits 1 to 9 at most one time each, place a digit in each box to create a true statement.

- **Grades 6–8:** Start by crowdsourcing an example of how students might start either of the examples provided here. As a group, try to fill out all the digits together and then let students give it another go on their own or in randomized groups of three!

- **Grades 9–12:** Ask students to show their thinking and draw attention to the different approaches that were taken. You might consider discussing whether or not an algorithmic approach to tackling this task would have worked!

HOT TIP

For more math tasks that help measure student progress from a growth perspective, check out the following:

- **Low-floor–high-ceiling tasks**: These are math tasks structured in a way that allows everyone to begin and work on the task at their own level of engagement; you can find these anywhere online.

- **Struggly**: An app full of math games and activities, created by Jo Boaler (www.struggly.com).

- **youcubed**: A website packed with math tasks and downloadable lessons, created by Jo Boaler and Cathy Williams (www.youcubed.org).

 Measure *Activity 2: Wrong Alert!*

Why

When the whole point is to be wrong . . . it's easy to be right! By asking students to name an answer to a question that they know is wrong, every student is given permission to feel successful in their own way.

The bonus of this Activity is that kids feel much more comfortable talking about why they think a math answer is wrong than why they think it's right. This Activity plays into any insecurities a student might have about their math abilities. It's also a great way to get an idea of what your students are thinking and to get them *communicating* that thinking in a way that feels nonthreatening!

When

Keep this Activity in your back pocket for any time students seem stumped about a concept or are scared of being wrong!

How

1. Make wrong the new right! The next time you're talking through how to tackle a problem, before you ask your students to solve the problem or to find the right answer, pause and ask the class, as a group, "What is a possible answer that would FOR SURE be wrong?"

   For example, let's say you're teaching estimation and asking students to estimate how many jelly beans are in a jar. Before you ask them to guess the correct amount, ask, "What is a number that is definitely the wrong answer?" You will find that kids are super excited to yell out all sorts of incorrect answers, not realizing that you've tricked them . . . into doing math!

2. Get a math talk going! Now that everyone's all hopped up on wrong answers, ask students to explain their thinking. This is the perfect way to get them engaged in conversations about math, and to let them know that they're slaying the game. EVERY student can feel "good at math" during this Activity because every wrong answer is technically right! It's a great opportunity to highlight each student's thinking, celebrate how great their "wrong" answer is, and to empower them to feel useful and helpful in a group!

Grade-Level Modifications

- **Grades K–2:** This is a great Activity to use when introducing concepts like number sense, estimation, and arithmetic!

- **Grades 3–5:** Use this activity when learning about probability, patterns, and odds!

- **Grades 6–8:** This Activity works especially well when discussing fractions, measurement, and percentages!

- **Grades 9–12:** Try this Activity when teaching anything related to slope, geometry, or any type of function!

Measure *Activity 3: Which One Doesn't Belong?*

Why

As we covered in the previous Activity, when there's no official "right answer," students are more relaxed and eager to participate in mathematical activities and to feel a sense of accomplishment as a result. Kids feel challenged to think critically and to explain their reasoning in a way that isn't intimidating. (Look for the free download on the Math Therapy website.)

I have done this Activity so many times with both kids AND adults, and every time, it leads to the deep math talks that we educators go to sleep dreaming of! The ask is simple: Which one doesn't belong? And the catch? There IS no right answer! Once I showed a group of educators photos of four different celebrities and famous groups: the Spice Girls, Beyoncé, NSYNC, and Keanu Reeves. I asked them which one didn't belong. The group erupted in MATHEMATICAL MAYHEM! Some of the answers I got included these:

- Keanu Reeves doesn't belong *because he's the only guy.*

- Keanu Reeves doesn't belong *because he's Vanessa's soulmate.*

- Beyoncé doesn't belong *because she's the only solo female.*

- The Spice Girls don't belong *because they're the only non-Americans.* (FYI—this is incorrect because what most people don't know . . . except for his die-hard fans aka ME . . . is that Keanu is a PROUD Canadian!)

It went on and on—and generated SUCH good discussion! Every response was met with, "Ohhhhh, I hadn't thought of THAT!" We were discussing patterns, quantities, attributes, and all sorts of math-y things—all while arguing over A-list celebs. So . . . pretty much a dream come true for me, and it likely will be for your students too!

When

Use this Activity any time! It's your best-kept secret, especially with a group of kids who are afraid to take risks in math class!

How

1. Pick your fab four! Display a set of four math visuals and ask students to individually choose and justify which one doesn't belong! Use your students' interests when picking the images! You can get crafty with it if there's a specific skill you're trying to build. For example, if you're working on shape properties, you could pick four different food items that vary in terms of shape and size (see Figure 7.3).

FIGURE 7.3 A SUPER-SWEET DISPLAY OF SHAPES

SOURCE: Canva

2. Have at it! In groups of three, have students discuss which one doesn't belong. Let them know that they will need to come to an agreement of ONE choice to present to the class, so they're going to need to work together and share their reasoning for which one they're going to pick and defend to the whole class!

3. Share with the class! Each group will pick a spokesperson to present their top choice of which one doesn't belong to the class. Make sure you ask them to emphasize their reasoning and let them know how much you value their critical thinking. Give specific feedback about how much you appreciate their creativity and uniqueness in coming to their choice!

4. Reflect! Reflect on the diversity of perspectives in the class and discuss the importance of multiple solutions in mathematics. Summarize key concepts, emphasize critical thinking and varied viewpoints, and congratulate everyone's role in coming up with unique ideas by demonstrating mathematical reasoning!

Grade-Level Modifications

- **Grades K–2:** Buff up this activity by using real objects instead of images: stuffed animals, favorite toys, candy . . . get creative and ask your students to bring their fave items in!

- **Grades 3–5:** Consider using photos of candy, animals, popular cartoon characters, or snacks!

- **Grades 6–8:** You can get even more bang for your buck by using this Activity as a way to get to know your students! Survey them on their fave foods, pets, toys, or celebs and use those to inform what photos you select for this Activity.

- **Grades 9–12:** Take this Activity up a notch by involving whatever math concept it is that you're currently working on. For example, you might ask, "What fraction doesn't belong?" or "What shape doesn't belong?" This is a great way to coax students into taking risks around math in a low-stakes way!

MEASURE TOOLKIT: SIMPLE SWAPS

I know you're totally crushing it in your Math Therapy journey and that you're doing your best to try the Actions and Activities suggested through this book, but I also know that sometimes it feels like there just isn't time to try a whole new thing. These simple swaps will help you carry out Step 5 with zero prep time! Remember to do what you can and to start where you are and that it's the little things that often mean the most, so don't underestimate the effectiveness of the simple swaps you can make each and every day!

SWAP THIS	FOR THIS
Always being the one to set goals for your students	Involve them in the process of setting realistic and achievable math goals for themselves.
Relying solely on end-of-unit tests	Integrate frequent formative assessments like quizzes, exit tickets, or short exercises to measure progress.
Being the one "in charge of" measuring progress	Empower students to assess their own understanding by providing checklists, rubrics, or reflection prompts.
Being the only one to always initiate student meetings	Set up a system where students can take charge in booking a meeting with you to discuss their progress.
Vague grading criteria	Clearly communicate grading criteria and provide rubrics for assignments.

SWAP THIS	FOR THIS
Fixed benchmarks for all students	Encourage students to set personalized goals that reflect what they value in terms of individual areas of improvement and progress.
Generic feedback (e.g., "Good job!")	Provide personalized feedback that highlights specific improvements and areas of growth (e.g., "Great work on attempting a tough problem—you're getting so much better at taking risks!").
Relying solely on grades to measure progress	Consider growth-tracking tools that visually represent students' progress over time, allowing them to see their improvement.

 ## *MEASURE* TOOLKIT: THEY SAY, YOU SAY!

Here are some of the common statements we hear from students when they feel like they're not getting anywhere, along with—you guessed it—responses that might help them feel a teeny bit better!

WHEN THEY SAY . . .	YOU SAY . . .
"But I got a bad mark."	"Did you learn anything? Because that's what matters the most right now." (If it were me, I would personally then take this opportunity to remind everyone that I failed Grade 11 math . . . twice!)
"This is hard."	"We learn new things every day, and when they're hard, they actually help our brain grow. You don't have to love math, but think about it this way: If your brain is growing, it's going to help you in ALL areas of life, so hang onto that when the going gets tough!"
"Isn't the point to get the right answer?"	"No! Getting the right answer is *a* point, but it's not the only point. Other points include learning something, building a new skill, and so on."
"How can I tell if I'm successful if there are no marks?"	"What does success mean to you? Let's come up with a definition and see if you can feel successful regardless of seeing a mark on your page!"

⭐ *MEASURE* TOOLKIT: EXPANSION PACK

There is nothing I love more than helping my students see how far they've come. Back in Step 3 when we talked about motivation and success, one of the things I pointed out was that, because of the historical hierarchical nature of the classroom, students often feel like external validation is the only way for them to *feel good*. The great thing about the *Measure* Toolkit is that it's full of Actions and Activities to empower students to measure their OWN progress on their OWN terms. The more emphasis we place on their own self-assessments as valuable, and the more opportunities we give them TO self-assess, the more likely they are to become pros at internally validating themSELVES, both in our classrooms and beyond! Honestly, this is a skill I truly wish I had *for myself* and am working toward every day—it's hard when your whole life you've been taught to externally validate. Helping our students do this is a gift we can give them that will last a lifetime, for real. Because this step is so important and so rewarding, I've thrown in some more of my fave *Measure* Activities that you can deep dive into if you have time. You can find templates and instructions for most of these Activities by heading to the online resources on the Math Therapy website!

1. **Mindset Surveys:** Periodically administer mindset surveys to gauge students' beliefs about their abilities to learn math and keep track of their results so that you can visually show them how they have progressed in their math-adjacent skills! (You'll find a free download on the Math Therapy website.)

2. **Estimation Excitement:** Estimation activities are one of my fave ways to facilitate number talks, build number sense, and foster critical thinking skills in a way that feels low-stakes to students. Why? Because when we estimate, there's usually no one right answer right away! I'm going to use an example from Val Jansonius, a K–5 instructional coach in Iowa. These are the first four prompts from an estimation activity she uses with students. In total, there are 12 prompts for this Activity, and she encourages kids to make a new guess after each prompt, illustrating that with each additional piece of information, our minds can think critically about how that alters our initial estimate!

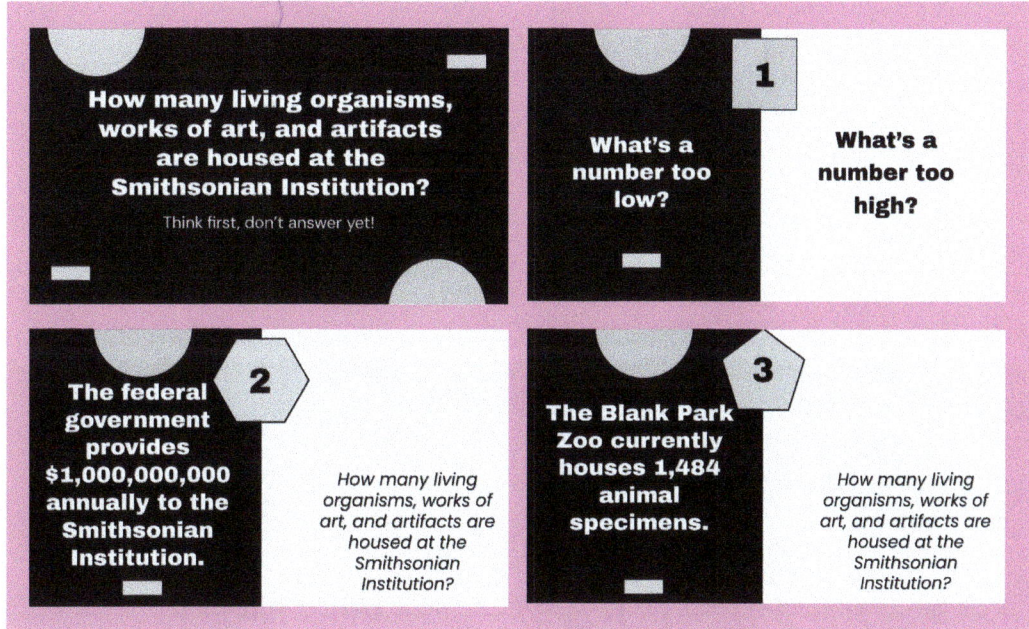

SOURCE: Used with permission of Val Jansonius

3. Conduct interviews with students about their learning processes. Set up one-on-one meetings with each of them to explore how they approach challenges, use resources, and demonstrate resilience in the face of difficulties.

HOT TIP

Want more bang for your Math Therapy buck? I have great news. As you've probably started to notice, many of the Actions, Activities, and simple swaps actually cover multiple steps of Math Therapy in one go. For example, something as simple as my suggestion to celebrate student success covers Step 1 (it shows students that there's more than one way to be good at math and highlights their progress), Step 4 (it provides them with a data point with which to rewrite their mathography), AND Step 5 (it provides a dopamine hit to motivate them to keep going *and* shows them that they have progressed)! HOW wild is that! So if you're feeling strapped for time, know that SO many of the strategies you take from this book will cover a LOT of ground in a SHORT amount of time! NOW GO GET 'EM!

STEP 5 TOOLKIT
TO GO
measure

measure
ACTIONS

1. BREAK THINGS INTO DOPAMINE BITES

Celebrate all the small wins that take place on the way toward learning a bigger concept!

2. CREATE A COMPLIMENT RESPONSE SYSTEM

Teach students how to respond to compliments in ways that allow them to celebrate.

3. TRACK UNIQUE STUDENT BENCHMARKS

Make students aware of all of the unique ways they build skills and show progress in your classroom.

measure
ACTIVITIES

1. OPEN MIDDLE MATH

Give kids autonomy to solve problems their own way using Open Middle Math!

2. WRONG ALERT!

Instead of asking students for the right answer, ask them to name ANY wrong answer!

3. WHICH ONE DOESN'T BELONG?

Engage students in rich math discussion by debating which item doesn't belong in a particular set!

measure
EXPANSION PACK

SMART GOAL SETUP ✳ MINDSET SURVEYS ✳ ESTIMATION EXCITEMENT

BEFORE WE MOVE ON

I'm sorry, did you just finish the FINAL STEP OF MATH THERAPY?! CLAP CLAP CLAP. I hate to see you go, but I love to watch you leave . . . Let's wrap up this final step with a treat and a moment of reflection, and then OFF TO THE FINAL CHAPTER!

Treat Yourself

Holy smokes, you have made it to the end of Step 5—the final step baby! Time to treat yourself by celebrating YOUR progress. Here are some of my fave ways to celebrate—pick one that speaks to you and DO IT! Options include: bake a cake, order a pizza, pour yourself a giant glass of bubbles (if you're into non-alc, my favorite faux-bubbles are Gruvi's DrySecco), take yourself out to dinner, get a new tattoo, buy yourself a piece of sparkly jewelry, look in the mirror and yell "YOU ROCK," or hey, go ahead and POP A CONFETTI CANNON!!!!!

ASK YOURSELF

1. Did this chapter change or challenge the way you have historically measured or celebrated success in your classroom? How?

2. How have you traditionally responded to students who get upset if they don't get an A+ on a test? Will you adjust the way you respond now? How?

3. What is one new way to celebrate student progress that you're going to try moving forward?

4. Which Action or Activity are you most excited to try with your students?

5. What is something in this chapter that makes you go, "aHA, this is totally going to make a difference"? Why?

6. What is something in this chapter that makes you go, "That will NEVER work with my students!" Why?

7. What are some of the challenges you may face when trying to implement the suggestions in this chapter? How might you prepare for these challenges ahead of time?

☮♡π

CHAPTER 8

SO NOW WHAT?

In this chapter, we get to:

1. Celebrate: We made it!

You did it. You have completed all 5 steps of Math Therapy! How do you feel?!

I feel like I need to make a grand gesture or some epic concluding statement, but as I've mentioned dozens of times throughout these pages, your Math Therapy journey has truly just begun—there is no finite end—because building a better relationship with math is all about guiding your students along a journey, not shuttling them off to a final destination.

We have all likely had many relationships in our lives and can attest to the fact that those most precious to us are an ever-evolving work in progress. I think of relationships as plants (a rich metaphor since I legit can't keep a plant alive). You can't just water a plant a bunch and be like, "Okay, there, done." In fact, that's probably why all my plants die . . . but never mind. The point is, when we get complacent in our relationships, they stagnate and wither away. We must consistently tend to them so they truly flourish. Our students' relationships with math are no different, and that's why there is no "end" to Math Therapy, only a series of new beginnings. All that being said, this book

> **Building a better relationship with math is all about guiding your students along a journey, not shuttling them off to a final destination.**

DOES have to have an actual ENDING, and I can think of no better way to end this book than by finally circling back to how I started it.

Remember how I told you that after my "I'm not a math person" limiting belief had been busted, my first thought was that I wanted to start a band? Well, I did. And here's what happened.

KEEP THE FAITH

So let me begin with the caveat that I had no evidence WHATSOEVER that I had any sort of "musical talent," by the way. In fact, much like my career as a math student, my "career" as a rockstar-wannabe was riddled with failures. For example, in my high school band, even though I wanted to be a drummer, I was handed the chimes as my assigned instrument. I don't know if you remember those from your high school days, but the chimes are this GIANT person-sized instrument that you hit with a hammer. Shortly after being handed the task of chimesmaster, during a huge competition I tripped and knocked the entire tower over, rendering our school band disqualified. Not a shining moment for anyone, especially me.

Then there was the time I auditioned for *Canadian Idol* (think: *American Idol* but, like, way less exciting). I stood outside, waiting in line in the blazing sun for 8 hours before my name was called, sang literally THREE words from "I'll Never Break Your Heart" by the Backstreet Boys, and was told in no uncertain terms to *go home*. But let me tell you, after my 96% in Ewa's math class, THE WORLD WAS MY OYSTER. I was no longer scared of failure, AT ALL.

One night in 2009, I headed to some dive bar to see a friend's band. I guess I had a little too much/just the right amount of wine, and the band was like, FINE, but they weren't great. At the end of their set, I marched right up to the stage, waved the singer over to where I was, and hollered, "YOU GUYS WERE GREAT, BUT YOU WOULD BE WAY BETTER WITH FEMALE HARMONIES," as though I was like, a label-exec who knew exactly what the industry demanded from budding young artists these days. I don't remember quite what happened after that or, like, WHY he didn't tell me to legit F-off, but the next day, my phone rang, and it was him. The singer of the band. (I guess I gave him my number?!??!)

As luck would have it, it didn't actually matter that I legit sucked. The band members had decided that none of them actually wanted to sing anymore and that never mind the whole female harmony suggestion I had approached them with the other night: They wanted a new lead singer, period. And it was going to be me.

Before I could so much as digest the fact that I had it's-Britney-B*tch-ed my way into an ACTUAL band, I was writing new lyrics to their old songs, we were rehearsing every single week, AND we had booked our first actual gig. The band was going to play a show. In front of actual, live people. AND I WAS THE FRONTWOMAN, GUYS.

Our first show ever was in the back room of a dive bar called Clinton's Tavern, where I had seen so many of my favorite bands over the years. A Toronto classic. It felt like anyone I had ever known showed up that night. In the weeks preceding the show, I started receiving messages from people I hadn't spoken to in forever, being like, "Wait . . . YOU'RE in a band?! I have GOT to see this," and "Haha, wait what?! Are you in a Britney Spears cover band or . . . ?!" That was the vibe, guys. Like, astonishment. Skepticism. Mild amusement at best. Story of my life, tbh.

Now, it's not like anything catastrophic happened. Like, the band played. I sang. I was wearing jeans, and I think I had one hand in my pocket the whole time like a total dork. Or like Alanis. (High-five if you got that reference.) I don't think I really even moved—I just focused on singing the right notes and, like, not totally messing the words up. People clapped politely. Clinked glasses. Congratulated me for ACTUALLY starting a band. Commented on how hilarious it was that all of a sudden I was fronting a band given that I had shown little to no musical aptitude (ever), aside from being obsessed with Britney Spears. Everyone was supportive. It was a good night. I had done it. Against all odds, there I was: the lead singer of a BAND.

And then I went outside for a smoke.

There I was, feeling like hot sh*t, when this guy came up to me and said, "Hey, are you the singer of that band?" It was happening. I was getting recognized; the paparazzi were sure to follow. I flipped my hair, flicked my cigarette, and faux-humbly said yes. And then, without an ounce of emotion, the guy goes, "Huh. You truly might be the worst singer I have EVER heard in my life."

(I refuse to apologize for my emoji usage because COME ON you know as well as I do that there was NO other way to end this story.)

Despite being straight-up RUDE, this random dude had given me a gift that night. He had challenged me to ask myself whether I was going to allow someone else to tell me how big my dreams could be, or whether I would make my dreams MINE and do whatever it took to reach them. I would be lying if I said it hadn't stung—like, a lot. But I'm being totally honest when

I tell you that not for ONE second did I consider quitting. My entire experience failing math almost THREE times and then overcoming my limiting beliefs had taught me that failure was an inevitable part of redefining success, so I wasn't scared off by the thought that I wasn't exactly where I wanted to be. If I was the worst singer this guy had ever heard, well, maybe I had work to do. So I hired a vocal coach.

The band broke up shortly after, and I started my very own. We called ourselves Goodnight Sunrise. We got good. I honed my inner Britney. I learned how to sing. I became the band's manager. I learned how to write songs, book shows, and ignore the haters. We toured the country, slept on disgusting couches, and got paid in beer. Eight years and 300 crappy bar shows later, we got a phone call that would remind us that anything truly IS possible.

Okay, I need you to pause right now because I want you to experience EXACTLY how I felt at what I STILL consider to be one of the most magical moments OF MY ENTIRE LIFE. Picture this: It's 8 a.m. on a Wednesday, and you're in a coffee shop the morning after you played a rock show. You went to bed at 3 a.m. Now, you're standing in line with David, your bff/ex-boyfriend/bandmate who happens to ALWAYS have their phone on silent; you're about to order a triple-shot latte (because: 8 a.m.!); there's a lineup behind you, and everyone is groggy and on their way to work and not feeling the vibe and suddenly—David's phone beeps. Why? Because as usual, his phone is on silent. The beep is letting him know that he has—get this—a VOICEMAIL (like who even leaves voicemails anymore?!). David pauses to check the message, you're ordering the lattes, and suddenly, you notice that David has gone pale. David finally holds the phone up to your ear so you can hear whatever this voicemail is. Are you ready? Here's what comes out of the speaker:

> Hello, Goodnight, Sunrise. This is Q107 radio calling to let you know that we loved the song you submitted and have selected your band to open for Bon Jovi at The Air Canada Centre [currently known as the Scotiabank Arena] this Saturday. Bye.

Can. You. IMAGINE??

Are *you* freaking out? *I* was!

Guys, no joke, I started running around the coffee shop *screaming*, "OUR BAND IS OPENING FOR BON JOVI! OUR BAND IS OPENING FOR BON JOVI!"

At that moment, everything just made sense. Like, all of the lessons I had learned over my lifetime had culminated in this one moment, and I felt like *anything was possible* (see Figure 8.1).

• You guys! This is MY band, Goodnight Sunrise, opening for Bon Jovi in front of 15,000 people at Toronto's Scotiabank Arena on May 12, 2018.

SOURCE: Ryan Brough

Okay, look, I know I said that phone call changed our lives forever blah blah—and it did, but not in the way you might think it did. It's not like all of a sudden we were rich and famous celebs or something. If you've learned ANYTHING in this book, it's that success isn't linear and that it never quite looks the way we expect it to. It changed our lives because it reaffirmed everything we had been taught: Practice makes progress, progress leads to growth, and growth takes you further than you were when you started, period. Wherever you go, there you are. But you won't get anywhere if you don't at least make a commitment to start the journey.

> *If you've learned ANYTHING in this book, it's that success isn't linear and that it never quite looks the way we expect it to.*

Five years later, no one would sign us to a label, so we started our own. We called it Rejection Records, and it's a nod to the fact that 95% of the time, we hear "no" from festivals, venues, music supervisors, booking agents—you name it. But all of that practice hearing "no" has put us in an incredible position of preparedness for the small but mighty percentage of situations where we get a "HECK, YES!"

Deepak Chopra (The Chopra Center, n.d.) claims that "luck is a concept invented and used by those who have not yet discovered the incredible power of living in alignment with the soul-spirit," but he ALSO says that "good luck is opportunity meeting preparedness."

Personally, I believe that both of these statements are equally true—and equally useful.

After our whole Bon Jovi situation, people would constantly come up to us and be all, "OMG, you guys are SO LUCKY that happened to you . . . lalalala!" and that response actually really bothered me.

Life isn't something that happens TO you. It happens BECAUSE of you. It wasn't just some random stroke of luck that we had been on THAT stage, THAT night. We weren't actually just livin' on a prayer—we were living out our dreams because we had worked hard to make them come true. None of these things had happened TO us. They had happened BECAUSE of us.

> *To this day, I wonder where I would be if I had never met the math teacher who changed my life. What would my relationship with math be like? More importantly, what would my relationship to hope and possibility be like?*

To this day, I wonder where I would be if I had never met Ewa Kasinska, the math teacher who changed my life. What would my relationship with math be like? More importantly, what would my relationship to hope and possibility be like? There is no way of knowing, but I can confidently say that it is the culmination of lessons I learned in that math class that took me far beyond my high school walls, all the way to that Scotiabank Arena stage.

WE'LL MAKE IT, I SWEAR

I recently visited a virtual school to carry out Math Therapy PD with their teachers. It was amazing to see how creatively these educators used their online platforms to engage students in innovative ways, and together, we discussed how we might use tools and strategies to bring Math Therapy to all of their students virtually. We all now know how hard it can be to really get to know our students virtually, but the good news is that it CAN be done and that by doing what we can to help our students truly feel seen, we can make a huge difference in how they heal from math trauma and build better relationships with math!

Who knows what the future holds, but making education accessible to more students will sometimes require us to flip over to delivering content virtually. That's why everything you have learned in this book can be modified to suit a virtual setting—and while it might require some creativity and flexibility,

I know that's something that ALL educators are absolute pros at, so I'm not worried AT ALL. If you currently teach in a virtual setting, don't be afraid to try the Actions and Activities in this book. Keep in mind that some stuff might be easier to modify than others and just start wherever you feel comfortable. The most important thing here is that the messaging behind the Math Therapy philosophy remains at the forefront of your classroom practice. Make sure every student knows that math trauma is a real thing, that they CAN build a better relationship with math, and that you're there to support them through it all! I promise that even just that alone will make a huge difference. Remember, all students just want to know that you believe in them, so whatever you do, make sure they know that you believe they'll make it (I swear)!

WHOA, YOU'RE HALFWAY THERE

Just as it's hard for our students to see how far they've come without you there to show them (throwback to Step 5!), I know it can be hard for educators to really see the impact they're making. If my Bon Jovi story illustrates anything, it's that success is an accumulation of teeny, tiny steps that often don't FEEL like progress in the moment. Sometimes progress feels like you're standing still, and sometimes it can feel like you're actually moving *backward* instead of forward, but I promise you that you're not.

As we close out the final few pages of our time together, I want to leave you with some takeaways that I have learned over years of working with incredible educators who have bravely taken on Math Therapy with their own students. Enjoy these juicy tidbits; I can't wait for you to eventually send me YOURS—my door is always open (metaphorically, of course—like, pls don't physically stalk me 🙂). I have set up my top-five takeaways as mathfirmations for you to repeat any time you need a reminder that, wherever you are on your Math Therapy journey, *it is exactly where you are meant to be.*

1. I am doing my best.

You are amazing, and you are doing the best you can with what you have, and it's important to remember that you don't exist in a vacuum! While we do have some control over what happens in our classrooms, we can't control what happens to our students when the bell rings. The consumption of anti-math media, mostly well-intentioned family members who insist math ability is genetic, and systemic factors that make it hard for our most vulnerable students to thrive, are just some examples of what we are contending with. And guess what? There's nothing you can do about any of them, but that doesn't make what you do in your classroom any less important or impactful. Accept what you can't control, focus on what you can, and just do your best!

2. I can do hard things.

Change is hard. Change is scary. But things don't change *if things don't change*! If you picked up this book, it's likely because something in you is igniting your curiosity for something different. Listen to that voice and let it fuel you when it seems like it might just be easier to stick with what you've been doing. Familiarity masks itself as security, but it is the fear of the unknown that keeps us seeking security in the first place. We don't need it. It is okay to feel untethered as you find your way along this new journey. Most of the amazing things you have in your life right now are a result of you doing hard things. Feel the fear, and *do it anyway!*

3. I am exactly where I am meant to be.

New things take time to figure out, but as Aerosmith once said, "Life's a journey, not a destination" (Tyler & Supa, 1993). There are going to be some scrapes and stumbles on your Math Therapy journey, and *that is okay!* Remember what I said at the beginning of this book? Learning to walk is a skill that most of us have, but it takes a while for most of us to find our stride! Don't worry about how long it takes you to get there. Just remember that you are making progress with EVERY single step you take. Keep your eyes on the prize and keep that forward momentum going—I know you'll figure out how to crush Math Therapy in your classroom!

4. I am starlight.

Don't try to be someone you're not—the magic you bring to Math Therapy is the secret ingredient that makes it work with YOUR students! Think of all of the incredible music there is out there. As I'm writing this, it is estimated that there are 11 MILLION artists on Spotify (Ruby, 2022). *11 MILLION!!!!* And not ONE of them has written the same song. Spotify's 574 million active listeners are scattered between those 11 million artists in every which way and would likely fight to the death that THEIR fave band is THE best band, no questions asked. Their uniqueness is what makes them special and beloved, and the same is true for the way YOU carry out Math Therapy as a part of your practice. You are the magic; you are the starlight—don't forget that and don't ever change!

5. So let them!

As I've learned over the past few years, whenever you're on the brink of success, there will ALWAYS be those who don't applaud you, or worse, who straight up want to see you fail. Call them haters, call them skeptics, call them old fashioned—but whatever you do—don't let them drown out your light.

If you're on social media, I don't need to tell you that our different opinions on what math education should look like can be divisive to the point of detriment, disrespect, and discordance. Don't let the noise drown out the sweet sounds of math relationships being healed in your classroom, don't let the status quo cloud your judgment, and don't forget that it is *not your job to please everyone* (and even if it was, it wouldn't even be POSSIBLE). Not everyone will agree with your choice to focus on healing math trauma and helping your students build better relationships with math, and people will talk smack about your belief that there is no such thing as "math person." And you know what Mel Robbins (2023b) would say? LET THEM!

6. I'm still here because I'm not there yet.

In 2021, my band was in the process of writing songs for our album *Against All Odds*. I was feeling pretty defeated and demoralized because, guys, let me tell you, the music industry has totally gone to sh*t (a discussion which is clearly beyond the scope of this book but one which I'm happy to rant about any time), and I literally thought I might pop my top if ONE more person told us that we were never going to get anywhere because we weren't already there *yet*. So in a rage of writer's block, suddenly an entire song came out of me, and the line that became the hook went like this: "I'm still here because I'm not there *yet*" (Kochberg & Vakharia, 2021). It's called "Wait for It," and it ended up being the most-streamed song on the album.

The thing is, just because you're not where you want to be doesn't mean you'll never get there. And in fact, you might just find that the "there" you wanted to reach changes over time, and that's okay—in fact, that's a sure sign of growth, so congrats, babe! I may have ultimately started a band and opened for Bon Jovi, but the "there" I imagined I would be looks nothing like the "here" I'm at now. No record label will sign us, most festivals won't book us without a label, and the costs of touring and making music keep getting higher while the potential for any sort of revenue stream keeps getting lower. But we're still out here, and we're still trying, and we're still having the best time—it just looks different than the way I once imagined it! We're still *here* because we're not there *yet*, and that's okay!

And guys, I know what you're all thinking: What about my whole "my dream is to marry Keanu Reeves" situation?! Well, as of the writing of this book, he IS engaged (to someone who actually seems quite lovely but also happens to not be me). But that's okay. I mean, maybe we WILL get married someday. Or maybe we'll just be soulmate besties. Or maybe my band will open for his band now that they've reunited. Maybe the "there" I envisioned once upon a time is now somewhere else, somewhere more aligned with the person I am becoming. This is all still tbd because we haven't met (yet). MY POINT IS THIS: The next time you feel like you're not far enough along on your journey, just repeat the following mantra—I'm still here because I'm not there YET. (And then feel free to yell SO JUST WAIT FOR IT at the top of your lungs!!!)

WHAT WILL *YOU* DO WITH IT?

So there you have it. My final words of encouragement. But as I've said repeatedly, your Math Therapy journey isn't over: It has just begun. You are undoing a *lifetime* of limiting beliefs around math, and it will take time. There will be relapses, but you have your trusty toolkits at your disposal any time you need them—you are prepared. You are ready. And yes, I designed those toolkits for you to use with your students, but pssst: You can use them *with yourself* too! Make sure YOU are taking the time to *Mythbust* your own expectations around what success looks like, to *Moderate* any moments of math trauma with mindfulness, *Motivate* yourself to keep going when the going gets tough, *Makeover* your story as a math educator to incorporate how far you've come, and above all, to *Measure*—AND CELEBRATE—all of the progress you're making in your classroom EVERY SINGLE DAY! You are making a difference. You are a beacon of hope. You are amazing.

> *Make sure YOU are taking the time to Mythbust your own expectations around what success looks like, to Moderate any moments of math trauma with mindfulness, Motivate yourself to keep going when the going gets tough, Makeover your story as a math educator to incorporate how far you've come, and above all, to Measure—AND CELEBRATE—all of the progress you're making in your classroom EVERY SINGLE DAY!*

At the beginning of this book, I shared my belief that math educators, more than anyone else on this planet, are in an incredibly unique position to impact young people's lives far beyond their classroom walls. I hope you now believe me. I hope that you can see how lucky your students are to have an adult in their lives who truly believes in their capacity to grow, dream, and change both themselves and the world in which we live. You have an opportunity that most people will never have. **What will you do with it?!**

what will you do with it?!

APPENDIX

START A MATH THERAPY BOOK CLUB!

Throughout this book, I have dropped sidebar "In The Moment" reflection questions that I invite you to turn into sizzling hot book club questions for your teacher friends. In the case that you want something a little more structured, here are 21 juicy questions to get your book club started. Feel free to make up your own, and if you do, send them to me—I want to see them! Happy Math Therapy-ing!

1. Which part of the book resonated with you the most, and why?

2. When you first picked up this book, did you think that you had any personal math trauma? Has your answer changed now that you've read the book? Explain!

3. How did the author's personal anecdotes or stories impact your understanding of math trauma?

4. Which step of Math Therapy are you most excited about, and why?

5. Did any of the strategies mentioned in the book surprise you? Which ones do you think would be most effective in your classroom? Which ones do you think you would have a hard time with?

6. Think back to your own experiences with math education. Can you relate to any of the struggles or challenges discussed in the book?

7. How do you think societal attitudes toward math contribute to students' math trauma? What do you think you personally can do to combat these attitudes that students might enter your classroom with?

8. The book talks about the importance of building positive relationships with math. How much of a priority do you think this should be for educators, given everything else they're expected to do with and for their students?

9. Were there any specific examples or case studies that stood out to you as particularly inspiring or insightful?

10. The author emphasizes the importance of making math relevant and relatable to students' lives. Can you think of any creative ways to accomplish this in a classroom?

11. What advice from the book would you give to someone who struggles with math anxiety?

12. The book discusses the connection between math confidence and confidence in other areas of life. Do you agree with this argument? Why or why not?

13. Do you have any limiting beliefs in your own life (related or unrelated to math)? Do you feel like anything in this book has made you reconsider them? Explain!

14. The book explores the idea of reframing mistakes as opportunities for learning and growth. Can you share an experience where you've personally benefited from embracing this mindset?

15. Do you think the cultural perception of math as a "hard" subject contributes to math anxiety? Do you think this perception is justified? How can we challenge and change this perception?

16. One of the major themes of Math Therapy is raising the stakes in math class so that *all* students see value in the math-adjacent skills they're building in your classroom. Can you think of an example of when a math-adjacent skill you learned in math class helped you achieve something in your actual life?

17. The author suggests incorporating mindfulness practices into math education. Do you think this could be beneficial? Why or why not?

18. The book talks about the importance of creating a supportive learning environment but warns that a truly "safe" space can never be promised to students. Do you agree? Why or why not? What specific actions can teachers take to foster a supportive space in their classrooms?

19. Reflecting on your own experiences with math, do you think you've ever unintentionally contributed to perpetuating math trauma in others? How can we break this cycle?

20. A key part of Step 5 of Math Therapy is redefining "success." How did you define success before picking up this book? How do you define it now? Consider both success for *yourself* personally as well as for the students in your classroom.

21. What is the first thing you're going to do to begin to cultivate a Math Therapy classroom with your students on Monday?!

REFERENCES

Ackerman, C. (2019). 19 narrative therapy techniques, interventions + worksheets. *PositivePsychology.com*. https://positive psychology.com/narrative-therapy/

Bartlett, K. A., & Camba, J. D. (2023). Gender differences in spatial ability: A critical review. *Educational Psychology Review*, *35*(1), 8. https://link.springer.com/article/ 10.1007/s10648-023-09728-2

Bernstein, G. (2019). *Super attractor: Methods for manifesting a life beyond your wildest dreams*. Hay House.

Blue Zones. (n.d.). *Blue zones—Live longer, better*. https://www.bluezones.com

Boaler, J. (2013). Ability and mathematics: The mindset revolution that is reshaping education. *Forum*, *55*(1), 143–152.

Boaler, J. (2014). Research suggests that timed tests cause math anxiety. *Teaching Children Mathematics*, *20*(8), 469. https://doi.org/ 10.5951/teacchilmath.20.8.0469

Boaler, J. (2015, January 28). *Fluency without fear*. youcubed. https://www.youcubed.org/ evidence/fluency-without-fear/

Boaler, J. (2016). *Mathematical mindsets: Unleashing students' potential through creative math, inspiring messages and innovative teaching*. Jossey-Bass.

Boaler, J. (2019, May). Prove it to me! *Mathematics Teaching in the Middle School*, *24*(7), 422–428. https://doi.org/10.5951/ mathteacmiddscho.24.7.0422

Boaler, J. (2024). *Math-ish: Finding creativity, diversity, and meaning in mathematics*. HarperCollins.

Boaler, J., Dieckmann, J. A., LaMar, T., Leshin, M., Selbach-Allen, M., & Pérez-Núñez, G. (2021). The transformative impact of a mathematical mindset experience taught at scale. *Frontiers in Education*, *6*, 784393. https://doi.org/10.3389/feduc .2021.784393

Brunat, E. (n.d.). *Esther Brunat*. Retrieved January 2, 2024, from https://estherbrunat .com

Centers for Disease Control and Prevention. (2023, March 8). *Data and statistics on children's mental health*. CDC. https://www .cdc.gov/childrensmentalhealth/data.html

Cherry, K. (2023, November 13). What is the negativity bias? *Verywell Mind*. https://www.verywellmind.com/negative-bias-4589618#citation-4

Cho, H., Ryu, S., Noh, J., & Lee, J. (2016). The effectiveness of daily mindful breathing practices on test anxiety of students. *PLoS One*, *11*(10), e0164822. https://doi .org/10.1371/journal.pone.0164822

The Chopra Center. (n.d.). *The Chopra Center*. https://chopra.com

Clear, J. (2018). *Atomic habits*. Penguin.

Columbia Undergraduate Admissions. (n.d.). *Columbia announces class of 2027 admissions decisions*. https://undergrad .admissions.columbia.edu/columbia-announces-class-2027-admissions-decisions

Corbett, B., Weinberg, L., & Duarte, A. (2017). The effect of mild acute stress during memory consolidation on emotional recognition memory. *Neurobiology of Learning and Memory*, *145*, 34–44.

Crawley, M. (2023, June 21). *What Ontario's rising high school grades mean for university admissions*. CBC. https://www.cbc.ca/ news/canada/toronto/ontario-university-admission-rising-grades-1.6875357

Dobrow, D. (2022, March 20). *Kyocera VP-210: This was the first camera phone in history.* Techweekmag.com. https://www.techweek mag.com/news/mobile/kyocera-vp-210-this-was-the-first-camera-phone-in-history/

Doran, G. T. (1981). There's a S.M.A.R.T. way to write management's goals and objectives. *Management Review, 70,* 35–36.

Dowker, A., Sarkar, A., & Looi, C. Y. (2016). Mathematics anxiety: What have we learned in 60 years? *Frontiers in Psychology,* 7(508), 1–16. https://doi.org/10.3389/fpsyg .2016.00508

Duncombe, C. (2021). *Unequal opportunities: Fewer resources, worse outcomes for students in schools with concentrated poverty.* The Commonwealth Institute. https://thecommonwealthinstitute.org/ research/unequal-opportunities-fewer-resources-worse-outcomes-for-students-in-schools-with-concentrated-poverty/

Dweck, C. S. (2007). *Mindset: The new psychology of success.* Ballantine Books.

Ehsaei, P. (n.d.). Why the TikTok trend "girl math" regresses women to the 1970s. *Forbes.* https://www.forbes.com/sites/pattieeh saei/2023/08/29/why-the-tiktok-trend-girl-math-regresses-women-to-the-1970s/

Flannery, M. E. (n.d.). The epidemic of anxiety among today's students. *NEA.* https://www .nea.org/nea-today/all-news-articles/ epidemic-anxiety-among-todays-students

Gilreath, A. (2023, September 28). Teachers conquering their math anxiety. *The Hechinger Report.* https://hechingerreport.org/teachers-conquering-their-math-anxiety/

Gomez, L. E., & Bernet, P. (2019). Diversity improves performance and outcomes. *Journal of the National Medical Association,* 111(4), 383–392. https://doi.org/10.1016/j .jnma.2019.01.006

Goodyear, S. (2020, September 11). Mathematician comes to defence of TikTok teen blasted for saying math isn't real. *CBC.* https://www .cbc.ca/radio/asithappens/as-it-happens-friday-edition-1.5720730/mathematician-comes-to-defence-of-tiktok-teen-blasted-for-saying-math-isn-t-real-1.5720731

Gonzalez, L. (2023). *Bad at math?* Corwin.

Hachey, A. C. (2021). Advancing early STEM identity development: Insights into early childhood mathematics education. *Journal of Mathematics Education,* 14(2), 1–7. https://doi.org/10.26711/007577152 790070

Hamid, A. A., Pettibone, J. R., Mabrouk, O. S., Hetrick, V. L., Schmidt, R., Vander Weele, C. M., Kennedy, R. T., Aragona, B. J., & Berke, J. D. (2015). Mesolimbic dopamine signals the value of work. *Nature Neuroscience,* 19(1), 117–126. https://doi .org/10.1038/nn.4173

Hamkins, S. (2014). *The art of narrative psychiatry.* Oxford University Press.

Harackiewicz, J. M., Rozek, C. S., Hulleman, C. S., & Hyde, J. S. (2012, August 1). Helping parents to motivate adolescents in mathematics and science: An experimental test of a utility-value intervention. *Psychological Science,* 23(8), 899–906. https://doi.org/ 10.1177/0956797611435530

Headspace. (2017). *Meditation and sleep made simple.* https://www.headspace.com.

Herndon, J. (2018, May 30). *What is a self-serving bias and what are some examples of it?* Healthline Media. https://www.health line.com/health/self-serving-bias

Heston, J. (2015). *The unlimited self.* Createspace.

Hoffower, H. (2018, September 24). 9 ways college is different for millennials than it was for previous generations. *Business Insider.* https://www.businessinsider.com/how-college-is-different-now-then-millennials-vs-baby-boomers-2018-9

Hottinger, S. N. (2017). *Inventing the mathematician: Gender, race, and our cultural understanding of mathematics.* State University of New York Press.

Hunt, K. (2019, November 8). Brain scans don't lie: The minds of girls and boys are equal in math. *CNN.* https://www .cnn.com/2019/11/08/health/math-boys-girls-brains-scn/index.html

Hurley, J. (2023, August 27). Six graduates from the same Toronto high school earned 100% grades. What's in the water at Father John Redmond? *Toronto Star.* https://www.thestar.com/news/gta/six-graduates-from-the-same-toronto-high-school-earned-100-grades-what-s-in-the/ article_00437d74-443b-5950-ba75-b95a501ed561.html

Kahneman, D. (2011). *Thinking, fast and slow.* Farrar, Straus and Giroux.

Kaplinsky, R. (2020). *Open middle math: Problems that unlock student thinking, grades 6–12.* Stenhouse.

Kochberg, D., & Vakharia, V. (2021). Wait for it [Song]. On *Against All Odds.* Rejection Records.

Kropf, N. P., & Tandy, C. (1998). Narrative therapy with older clients. *Clinical Gerontologist, 18*(4), 3–16. https://doi.org/10.1300/j018v18n04_02

Kubala, K., & Lebow, H. I. (2022, April 19). *How can childhood trauma affect learning and education?* Psych Central. https://psychcentral.com/ptsd/complex-ptsd-trauma-learning-and-behavior-in-the-classroom

Lambert, R. (2024). *Rethinking disability and mathematics.* Corwin.

Levine, S. C., & Pantoja, N. (2021). Development of children's math attitudes: Gender differences, key socializers, and intervention approaches. *Developmental Review, 62,* 100997. https://doi.org/10.1016/j.dr.2021.100997

Liljedahl, P. (2021). *Building thinking classrooms in mathematics, grades K–12.* Corwin.

Louie, N. L. (2017). The culture of exclusion in mathematics education and its persistence in equity-oriented teaching. *Journal for Research in Mathematics Education, 48*(5), 488. https://doi.org/10.5951/jresematheduc.48.5.0488

Love Is Blind (TV series). (2020, November 17). Wikipedia. https://en.wikipedia.org/wiki/Love_Is_Blind_(TV_series)

Luzniak, C. (2020). *Up for debate!: Exploring math through argument.* Stenhouse.

Maté, G., & Maté, D. (2022). *The myth of normal: Trauma, illness, & healing in a toxic culture.* Avery.

Matthews, L. E., Jones, S. M., & Parker, Y. A. (2022). *Engaging in culturally relevant math tasks, K–5: Fostering hope in the elementary classroom.* Corwin.

McConchie, L. (2022). *Motivating students with the brain in mind.* ASCD. https://www.ascd.org/el/articles/motivating-students-with-the-brain-in-mind

McDermott, N., & Spann, R. T. (2022, November 9). How gratitude can transform your mental health. *Forbes Health.* https://www.forbes.com/health/mind/mental-health-benefits-of-gratitude/

Mervosh, S., & Wu, A. (2022, October 24). Math scores fell in nearly every state, and reading dipped on national exam. *The New York Times.* https://www.nytimes.com/2022/10/24/us/math-reading-scores-pandemic.html

Mindshift. (2017, November 13). *Students share the downside of being labeled "gifted."* KQED. https://www.kqed.org/mindshift/49653/students-share-the-downside-of-being-labeled-gifted

Müller-Pinzler, L., Czekalla, N., Mayer, A. V., Stolz, D. S., Gazzola, V., Keysers, C., Paulus, F. M., & Krach, S. (2019). Negativity-bias in forming beliefs about own abilities. *Scientific Reports, 9*(1), 1–15. https://www.nature.com/articles/s41598-019-50821-w

National Institute of Mental Health. (2023). *Any anxiety disorder.* https://www.nimh.nih.gov/health/statistics/any-anxiety-disorder

National Science Foundation (NSF). (n.d.). *Home.* https://ncses.nsf.gov

Noonoo, S. (2023, April 21). 6 ways to merge literacy with mathematics. *Edutopia.* https://www.edutopia.org/article/6-ways-to-merge-literacy-with-mathematics/

Norris, C. J., Creem, D., Hendler, R., & Kober, H. (2018). Corrigendum: Brief mindfulness meditation improves attention in novices: Evidence from ERPs and moderation by neuroticism. *Frontiers in Human Neuroscience, 12,* 315. https://doi.org/10.3389/fnhum.2018.00342

Owen, S. E. (2021). An exploration of math trauma through ability grouping and teacher language in elementary schools. *Honors Projects.* 129. https://digitalcommons.spu.edu/honorsprojects/129

Oxford Dictionary. (2023). *Oxford languages.* Oxford University Press. https://languages.oup.com/google-dictionary-en/

Phillips, K. W. (2014). How diversity makes us smarter. *Scientific American, 311*(4), 42–47. https://www.scientificamerican.com/article/how-diversity-makes-us-smarter/

Pitman, K. (2023). How centuries of sexism excluded women from science—and how to redress the balance. *Nature, 619*(7969),

243–244. https://www.womeninscience.com/article/view?id=1241

PrepScholar. (2017). This year's University of Michigan admission requirements. *Prepscholar.com.* https://www.prepscholar.com/sat/s/colleges/University-of-Michigan-admission-requirements

Robbins, M. (2023a, December 3). *5 ways to improve your subconscious mind & be happier in 2024: Amazing insight from Dr. Paul Conti. The Mel Robbins Podcast.* Episode 125. https://www.melrobbins.com/podcasts/episode-125

Robbins, M. (2023b, May 28). The "let them" theory: A life-changing mindset hack that 15 million people can't stop talking about. *The Mel Robbins Podcast.* Episode 70. https://www.melrobbins.com/podcasts/episode-70

Rosser, E. (2010). Heroin chic: The fashion phenomenon. *Inquiry: Critical Thinking Across the Disciplines, 2*(12), 1–4.

Rothstein, L., & Stromme, D. (n.d.). *Two for you video series: RAS (reticular activating system).* University of Minnesota Extension. https://extension.umn.edu/two-you-video-series/ras

Ruby, D. (2022, January 29). *Spotify stats 2022: All data & stats listed.* Demandsage. https://www.demandsage.com/spotify-stats/

Simone, F. (2017, December 4). Negative self-talk: Don't let it overwhelm you. *Psychology Today Canada.* https://www.psychologytoday.com/ca/blog/family-affair/201712/negative-self-talk-dont-let-it-overwhelm-you

Skinner, B. F. (1963). Operant behavior. *American Psychologist, 18,* 503–515.

Sokolowski, H. M., & Ansari, D. (2017). Who is afraid of math? What is math anxiety? and What can you do about it? *Frontiers for Young Minds, 5.* https://doi.org/10.3389/frym.2017.00057

Sparks, S. D. (2022, June 3). 3 out of 4 gifted Black students never get identified. Here's how to find them. *Education Week.* https://www.edweek.org/teaching-learning/3-out-of-4-gifted-black-students-never-get-identified-heres-how-to-find-them/2022/06

Stanborough, R. J. (2020, February 4). *How to change negative thinking with cognitive restructuring.* Healthline. https://www.healthline.com/health/cognitive-restructuring

Strong, S., & Butterfield, G. (2022). *Dear math: Why kids hate math and what teachers can do about it.* Times 10.

Sun, K. L. (2014). Math can-do: Column. *USA TODAY.* https://www.usatoday.com/story/opinion/2014/07/09/math-misconceptions-education-reform-column/12430181/

Swift, T. (2022, October 21). *Taylor Swift—Anti-Hero* (official music video). www.youtube.com. https://www.youtube.com/watch?v=b1kbLwvqugk

Teasdale, J., Williams, M., & Segal, Z. V. (2014). *The mindful way workbook: An 8-week program to free yourself from depression and emotional distress.* Guilford Press.

Tyler, S., & Supa, R. (1993, November). *Amazing* [Review of *Amazing*]. Bruce Fairbairn.

The Understood Team. (2017). Learning disabilities by the numbers. *Understood.org.* https://www.understood.org/en/articles/learning-disabilities-by-the-numbers

Vakharia, V. (2010). *Peace, love and pi: Imagining a world where Paris Hilton loves mathematics* [Dissertation]. University of British Columbia, Vancouver, BC.

Vakharia, V. (Host). (2020, June 18). Recalculating your destiny w/ Miguel Escobar [Audio podcast episode]. In *Math Therapy.* https://www.maththerapypodcast.com/1734392/8155458-s2e03-recalculating-your-destiny-w-miguel-escobar

Vakharia, V. (Host). (2021, June 10). How math rehabilitated a murderer w/ Christopher Havens [Audio podcast episode]. In *Math Therapy.* https://www.maththerapypodcast.com/1734392/12868918-mathers-gonna-math-w-deborah-peart

Vakharia, V. (Host). (2023a, May 18). Mathers gonna math w/ Deborah Peart [Audio podcast episode]. In *Math Therapy.* https://www.maththerapypodcast.com/1734392/12868918-mathers-gonna-math-w-deborah-peart

Vakharia, V. (Host). (2023b, June 1). How trauma affects the brain w/ Liesl McConchie [Audio podcast episode]. In *Math Therapy.* https://www.maththerapypodcast.com/

1734392/12949358-how-trauma-affects-the-brain-w-liesl-mcconchie

Walton, G. M., & Cohen, G. L. (2007). A question of belonging: Race, social fit, and achievement. *Journal of Personality and Social Psychology, 92*(1), 82–96.

Walton, J. L. (2001). *The effect of grading the work of fourth grade students in red ink and their academic self-esteem* [Theses and Dissertations]. 1614. Rowan University, Glassboro, New Jersey. https://rdw.rowan.edu/cgi/viewcontent.cgi?article=2614&context=etd

Watson, C. M. (2024). *About—Crystal M. Watson*. https://crystalmwatson.com/about-crystal/

West, M. (2022, September 29). Math anxiety: Definition, symptoms, causes, and tips. *MedicalNewsToday*. https://www.medicalnewstoday.com/articles/math-anxiety-definition-symptoms-causes-and-tips#math-anxiety

INDEX

**JENNIFER M. BAY-WILLIAMS,
JOHN J. SANGIOVANNI, ROSALBA SERRANO,
SHERRI MARTINIE, JENNIFER SUH,
C. DAVID WALTERS, SUSIE KATT**

Because fluency is so much more than
basic facts and algorithms.

Grades K–8

**MARIA DEL ROSARIO
ZAVALA,
JULIA MARIA AGUIRRE**

Discover innovative equity-
based culturally responsive
mathematics instruction that
unlocks the mathematical
heart of each student.

Grades K–8

**IVANNIA SOTO,
THEODORE RUIZ SAGUN,
MICHAEL BEIERSDORF**

Focus on the literacy
opportunities that multilingual
students can achieve when
language scaffolds are
taught alongside rigorous
math and science content.

Grades K–8

**JOHN J. SANGIOVANNI, SUSIE KATT,
LATRENDA D. KNIGHTEN, GEORGINA RIVERA,
FREDERICK L. DILLON, AYANNA D. PERRY,
ANDREA CHENG, JENNIFER OUTZS, KAREN MESMER,
ENYA GRANDOS, KEVIN GANT, LAURA SHAFER**

Actionable answers to your most pressing
questions about teaching elementary math,
secondary math, and secondary science.

Elementary, Secondary

**CHRISTA JACKSON, KRISTIN L. COOK,
SARAH B. BUSH,
MARGARET MOHR-SCHROEDER,
CATHRINE MAIORCA, THOMAS ROBERTS**

Help educators create integrated STEM
learning experiences that are inclusive for all
students and allow them to experience STEM
as scientists, innovators, mathematicians,
creators, engineers, and technology experts!

Grades PreK–5 and Grades 6–12

CORWIN

A Sage Company

CORWIN HAS ONE MISSION: to enhance education through intentional professional learning.

We build long-term relationships with our authors, educators, clients, and associations who partner with us to develop and continuously improve the best evidence-based practices that establish and support lifelong learning.